Trade Liberalisation, Manufacturing Growth
and Employment in Bangladesh

This report is published by Academic Foundation, New Delhi under arrangement with the International Labour Organization (ILO), Geneva.

The report was prepared under a programme jointly implemented by the Centre for Policy Dialogue (CPD), Dhaka and the International Labour Organisation (ILO), Geneva

Report Prepared by:
Mustafizur Rahman, Executive Director, CPD, Dhaka
Wasel Bin Shadat, Senior Research Associate, CPD, Dhaka
Selim Raihan, Associate Professor, University of Dhaka

Research Assistance:
Suparna Hasan, Senior Research Associate, CPD, Dhaka
Sohani Fatehin, Programme Associate, CPD, Dhaka
Md. Shawkat Hossain Bipu, Intern, CPD, Dhaka

Trade Liberalisation, Manufacturing Growth and Employment in Bangladesh

Mustafizur Rahman
Wasel Bin Shadat
Selim Raihan

International Labour Office
Geneva

Academic Foundation
New Delhi

First published in 2010 by

Academic Foundation
4772/23 Bharat Ram Road, (23 Ansari Road), Darya Ganj,
New Delhi - 110002, India.
email: books@academicfoundation.com
www.academicfoundation.com

Trade Liberalisation, Manufacturing Growth and Employment in Bangladesh

ISBN 978-81-7188-819-1

Designed and typeset by Italics India, New Delhi
Cover design copyright: Academic Foundation
Printed and bound in India

Contents

List of Tables, Figures, Boxes, Annexes and Annex Tables

TABLES

FIGURES

BOXES

ANNEXES

ANNEX TABLES

List of Acronyms

BB	: Bangladesh Bank
BBS	: Bangladesh Bureau of Statistics
BEI	: Bangladesh Enterprise Institute
BER	: Bangladesh Economic Review
BSIC	: Bangladesh Standard Industrial Classification
BTC	: Bangladesh Tariff Commission
C-D	: Cobb Douglas
CD	: Custom Duty
CES	: Constant Elasticity of Substitution
CET	: Constant Elasticity of Transformation
CGE	: Computable General Equilibrium
CIF	: Cost, Insurance and Freight
CLF	: Civilian Labour Force
CMI	: Census of Manufacturing Industries
CMT	: Cut, Make and Trim
CPI	: Consumer Pricc Index
EER for Import	: Effective Exchange Rate for Import
EER for X	: Effective Exchange Rate for Export
EOR	: Export Orientation Ratio
EPB	: Export Promotion Bureau
EPZ	: Export Processing Zones
ERP	: Effective Rate of Protection

ESAF	: Enhanced Structural Adjustment Facility
FDI	: Foreign Direct Investment
FOB	: Free on Board
GATT	: General Agreement on Tariffs and Trade
GDP	: Gross Domestic Product
GSP	: Generalised System of Preferences
HOS	: Heckscher-Ohlin-Samuelson
HS Code	: Harmonised System Code
IFRI	: International Food Policy Research Institute
IMF	: International Monetary Fund
IPOs	: Import Policy Orders
ISIC	: International Standard of International Classification
L/C	: Letter of Credit
LDCs	: Least Developed Countries
LFS	: Labour Force Survey
LM Test	: Langrangian Multiplier Test
M	: Import
MFN	: Most-favoured-nation
NBR	: National Board of Revenue
NOE	: Number of Employee
NPW	: Non-Productive Worker
NTBs	: Non-Tariff Barriers
OECD	: Organisation for Economic Co-operation and Development
PW	: Productive Worker
QR	: Quantitative Restrictions
RD	: Regulatory Duty

REER	: Real Effective Exchange Rate
RMG	: Readymade Garments
SAF	: Structural Adjustment Facility
SAM	: Social Accounting Matrix
SAP	: Structural Adjustment policy
SD	: Supplementary Duty
SMEs	: Small and Medium Enterprises
SOE	: State Owned Enterprises
SST	: Stolper-Samuelson Theorem
TI	: Tariff Incidence
TOT	: Terms of Trade
UECM	: Unrestricted Error Correction Mechanism
UNCOMTRADE	: United Nations Commodity Trade
UW	: Unweighted
VA	: Value Added
VAFC	: Value Added Factor Cost
W	: Weighted
WB	: World Bank
WDI	: World Development Indicators
WPW	: Wage Per Worker
WTO	: World Trade Organization
X	: Export
XPB	: Export Performance Benefit
XPL	: Export Performance Licensing

1 Introduction

The purpose of the present study is to understand the nature of relationship between trade reforms and labour market dynamics in the context of Bangladesh economy to make an assessment about the impact of various measures of trade liberalisation on macro and sectoral disaggregated level movements in the labour market.

Over the recent past decades, an overwhelming majority of developing countries have been pursuing policy of trade liberalisation, albeit with different pace with varying degree of outcomes. These reforms in trade related areas included lowering of import tariffs, and para-tariffs (special duties, surcharges), removal of anti-export bias in the economy, dismantling of non-tariff barriers such export and import bans, quotas and voluntary export restrictions and in general, reduction in the effective rates of protection for the domestic producers.

Advocates of trade liberalisation argued that trade liberalisation, when coupled with appropriate domestic reforms and sound macroeconomic policies, could be a powerful engine in promoting sustained economic growth, employment augmentation and poverty reduction. Countries

---

This volume has emerged from a collaborative study undertaken by the International Labour Office (ILO), Geneva and the Centre for Policy Dialogue (CPD), Dhaka, Bangladesh. The authors would like to express their deep gratitude to Dr. Rizwanul Islam, recently retired Special Adviser on Growth, Employment and Poverty Reduction; Employment Sector, ILO, Geneva for his extremely valuable guidance in the course of the entire study.

The authors would like to sincerely thank Suparna Hasan, Senior Research Associate, CPD for her able research support and for coordinating the study. Contribution of Sohani Fatehin, Programme Associate, CPD and Md. Shawkat Hossain Bipu, Intern, CPD is also thankfully acknowledged. Special appreciation for Mr. A.H.M. Ashrafuzzaman, Senior System Analyst, CPD who helped in preparing the manuscript for publication. The authors would also like to register their deep appreciation to the participants of a dialogue organised by the CPD in Dhaka where an earlier draft of the study was discussed by a distinguished set of experts.

undertaking such reforms would benefit from enhanced investment and export that would create new employment opportunities leading to higher economic growth and sustained poverty alleviation. They argued that by relaxing the constraints to trade, trade liberalisation policies open up opportunities for small economies to go beyond the confines of limited domestic market. This in turn creates opportunities for accessing foreign direct investment, facilitates technology transfer, creates marketing network, and provides the much-needed managerial and technical skills. Possible channels through which trade liberalisation could bring such benefits to developing economies included improved resource allocation in line with the social marginal cost and benefits, access to better technologies, inputs and intermediate goods, and greater domestic competition (Dornbusch, 1992). Supporters of a more open trading regime point to the Ricardian theory of comparative advantage, economic efficiency, contribution of competition, innovation and learning. Ebrill *et al.* (1999) argued that welfare gains can be achieved through efficiency in production, international competitiveness and increase in the volume of trade, which are outcomes of trade reforms. The World Bank estimates that removing all trade barriers could increase global income by US$ 2.8 trillion and lift 320 million people out of poverty by 2015 (World Bank, 2002).

However, opinions have varied with regard to the actual experience with respect to the impact and implications of trade reforms on economies of developing countries and this has become a subject of intense controversy and lively debate.

The arguments put forward by opponents of free trade (or a fast pace of trade liberalisation) could be put under three broad groups: economic arguments (infant industry arguments, the optimum tariff argument and arguments related to external economies or correction of domestic market distortions), non-economic arguments (arguments favouring self-sufficiency, developing a minimum level of industrialisation) and non-arguments (those based on inadequate analysis or understanding of the underlying issues and often driven by passion that generate more heat than light). Developing countries have traditionally tended to take resort to a host of measures to regulate their international trade, through a host of measures that included taxes (e.g. tariff), quasi-tariffs and para-tariffs (e.g.

regulatory or supplementary duties, surcharges), trade barriers (e.g. subsidies and quotas) and other non-tariff barriers (NTBs). Such policy tools were perceived to serve dual purposes, providing protection for the local industries and at the same time generating revenue for the government exchequer. Other objectives included correction of market distortions, improving the terms of trade by favourably affecting world market prices, and improving the distribution of income at home (Khattry and Rao, 2002).

On the other hand, the economic arguments for protection have also been extensively reviewed in the literature. Johnson (1958, 1971) found that the only valid proposition for tariff protection is the one based on the optimum tariff argument; other arguments for protection are arguments for subsidy. He also argued that the use of tariff in other cases can lead to further distortions. These economists have argued that free trade is by and large better than no trade or less trade, although this may not be a perfect policy.

Many economists, conceding that there are benefits to be accrued from trade liberalisation, puts emphasis on the fact that trade liberalisation gives rise to adjustment costs due to redistribution of resources from one sector to the other, from non-tradables to the tradables. Consequently, in recent literature, issues of adjustment costs associated with liberalisation, which tends to be more immediate and more visible than the benefits (which could be medium-to-long term) has been gaining prominence. The debate has largely been shifted from protection *versus* free trade to pacing, phasing and sequencing of the liberalisation, and the likely impact that these may have on the performance of the economy as also on distribution of income and social welfare. Stiglitz and Charlton (2006: 3) for example point out in this context, ".... any 'gross' welfare gains from trade liberalisation must be balanced against its associated costs ..... whereas it may take decades for multilateral trade reform to deliver gains to developing countries, the adjustment costs are automatic and usually upfront."

A key variable in this discourse on the impact of trade liberalisation on the economies of developing countries is 'employment' since employment creation is seen as a critically important factor that determines the economic and welfare implications of trade reforms in the context of the developing countries. Thus, the issue of creation or non-creation (even destruction) of employment opportunities as a result of trade reforms has been put at the centre of the debate on the impact of the policy of trade liberalisation pursued by developing countries. The present study is an attempt to capture this impact in the context of the developing economy of Bangladesh.

1.1 Rationale, Research Questions and Objectives of the Study

The discussion in the previous section has underscored the point that trade liberalisation could have diverse impact in terms of employment in developing countries. The review indicates that for each developing country, the nature, direction and depth of these impacts would vary and will be different. There are also a number of other policies and issues, which are not directly treated as trade reform policies, rather they should be treated as accompanying policies of trade reform, which also could have significant employment and wage impact. Some of these effects could act in the same direction with trade liberalisation while others could act in opposite direction. Apart from the data problems, one of the major difficulties to assess the impact of trade liberalisation is to separate its impact from other policies or events which are also likely to influence the outcomes along with the shock originating from trade liberalisation. A list of factors will need to be considered and carefully examined to assess the actual impact of trade reforms on the economy of a particular country.

Though there are many studies which look at the implications of trade liberalisation on growth and poverty in the context of Bangladesh, there is a much felt need for a more in-depth analysis of the impact of trade liberalisation in Bangladesh on the changing pattern of employment. There is a general agreement that not all sectors tend to benefit when economies open up. Identifying a balanced policy, keeping in mind the interests of both winners and losers is critically important from the perspective of the policymaker. However, this task has remained difficult because of poor quality of data, lack of consistency and absence of real time data on employment, both at macro and at sectoral levels.

While some studies have looked at certain aspects of the trade-employment nexus, detailed studies, with analyses of sector specific data as well as data from successive labour force surveys (including the recent years), are not there. The present study is an attempt to address this gap.

1.2 Focus of Study

Thus, the present study is concerned with an examination of changes in the employment scenario of Bangladesh as a consequence of the policy of trade liberalisation pursued over the past years. The study will attempt to review the structural changes in the labour market consequent to the implementation of major trade reforms in Bangladesh. An attempt will be made to relate such changes to the various phases of trade liberalisation. More concretely, the study will try to capture the impact of trade liberalisation on employment by using a number of indicators and econometric models such as production function, labour demand function and computable general equilibrium (CGE) modelling. Based on these analyses, the report will try to put forward some policy recommendations.

1.3 Analytical Framework and Data

The study presents a critical review of a selected set of literature on the impact of trade liberalisation on employment and wage inequality which particularly focus on Bangladesh. In the analytical part of the study, a number of tariff, trade and other relevant indicators have been used to identify various phases, pace and sequences of trade liberalising initiatives implemented by successive governments in Bangladesh. Along with this, trade policies including export policies, import policies and successive industrial policies have also been examined to identify major changes in the trade policy regime of Bangladesh over the past years.

It is to be mentioned here that major source of employment data in Bangladesh are Labour Force Survey (LFS) and Census of Manufacturing Industries (CMI) reports published by the BBS. LFS data presents the structure and distribution of the civilian labour force in Bangladesh. This data was analysed to capture the changes in the employment scenario, over the various phases of trade liberalisation, at a macro level. The change in employment dynamics in the industrial sub-sectors was analysed in detail by taking advantage of the disaggregated CMI dataset. Apart from employment status, changes in key industrial indicators (for example, labour productivity, capital-output ratio, etc.), tariff protection and trade data were also analysed to track changes over the various phases of trade liberalisation in Bangladesh. An attempt has also been made, based on an analysis of the CMI data, to identify sectors that are winners and those that are losers in the same context.

Rigorous modelling exercise was carried out through panel estimates of the production function and labour demand function, with several specifications, and by augmenting the trade liberalisation variables.

Finally, a CGE modelling exercise was carried out to investigate the potential implications of further trade liberalisation under various possible scenarios.

Period of analyses has varied across the various exercises depending on data availability and the purpose of analysis. However, the time span has been clearly identified in all such cases.

1.4 Limitations and Distinctive Features of the Study

It is to be reiterated here that dearth of data and lack of adequate labour market data comparability has tended to discourage researchers to undertake in-depth examination of the nexus between trade policies and labour market dynamics in the context of Bangladesh. Although this study has attempted to address many of the difficulties in this respect, it cannot claim to be free from limitations and shortcomings. First of all, the study had to be conducted by taking cognisance of the serious concern and debate among the experts and researchers with regard to the quality of employment data available in Bangladesh—both at macro level employment data (as provided in LFS reports) and meso-level industrial employment data and other industrial indicators (as reported in CMI). The issue of data correction and comparability was dealt with utmost importance and will be discussed in more details subsequently. Secondly, there is debate with regard to technical and definitional aspects of appropriate indicators for measuring the extent and speed of trade liberalisation (and inversely, in estimating the actual extent of real protection). In the present study, various indicators have been used as measures of trade liberalisation simultaneously and the definitions and sources of data have been mentioned clearly either in the main text or in the annexes. Thirdly, a number of important areas which are impacted upon by trade liberalisation including environment and technology upgradation, role of R&D and firm's capacity to adapt new technology, management efficiency, skill and productivity of workers, worker's welfare, role of employer's federations, right of trade unions and their scope of intervention, were not covered by this study. Fourthly, no case study of particular sectors was carried out through in-depth field investigation, nor was any detailed micro level analysis from workers' perspective such as wage inequality was performed as part of this study. Limitations of time and resources did not permit such an exhaustive study, although study of the aforesaid aspects remains important.

It would perhaps be pertinent here to mention about a number of distinctive features of the present study. Firstly, exhaustive data related corrections were carried out while comparing the data in the various LFSs with a view to adjusting the differences in definitional and age criteria in successive LFSs which has thus far put severe limitations of the use of the data by researchers. Secondly, the study has made use of the LFS 2005 data which has only recently been (unofficially) released. Thirdly, econometric analyses (estimation of production function and labour demand function) were carried out at 4-digit disaggregate industry-level data which is expected to provide more insightful results compared to what would have been possible from an analysis of only the more aggregate level data. It needs to be recognised upfront that the quality of 4-digit level industrial data as far as consistency and reliability of the data are concerned, is not very good. One of the distinctive features of this study is that a considerable effort has been put in to verify and correct the CMI data at 4-digit level so that an analysis based on the data was possible to undertake. Finally, sectoral (again at 4-digit BSIC level) trade and tariff data were estimated through BSIC-ISIC and ISIC-HS correspondence, also a novel attempt.[1] This allowed use of these sectoral tariff variables as a proxy of the extent of trade liberalisation at disaggregated sectoral levels, for the first

1. The issue of data corrections and mapping will be discussed in more details in subsequent chapters.

time in case of Bangladesh, rather than using just a dummy variable to indicate phases of trade liberalisation or using tariff variables at a more aggregate level such as effective rate of protection (ERP). This helped to provide a more realistic picture of the extent of sectoral protection and consequently made econometric results more robust.

1.5 Structure of the Report

The rest of this report is organised in the following manner: Chapter 2 presents a review of literature that looks at the impact of trade liberalisation on employment, wage, inequality and poverty; both conceptual and empirical evidences are discussed here. In Chapter 3, evolution of trade liberalisation policies in Bangladesh are discussed; various phases, pace and sequencing of trade reforms are identified here with the help of relevant indicators grouped under three broad heads: tariff, trade and other policy related indicators. In Chapter 4, trends and dynamics of changes in the labour market of Bangladesh are presented in the context of the various trade reform phases; changes in terms of productivity and wage indices have been analysed from a macro perspective based on Labour Force Survey (LFS) data and their interface with trade reforms have been examined. The focus of Chapter 5 is an in-depth analyses of the dynamics of employment scenarios in the manufacturing sector over various phases of trade liberalisation period. Census of Manufacturing Industries (CMI) data at disaggregate level has been used for this purpose. This chapter also investigates the performance of the manufacturing sectors during the various phases of trade liberalisation with the use of a range of indicators. Based on CMI data and use of panel econometric techniques, Chapter 6 has estimated production function and labour demand function with a special focus on the estimation of impact of trade liberalisation. Chapter 7 captures possible employment implications of further liberalisation through a simulation exercise applying the CGE modelling tools. Several scenarios were built-in into the model to study the possible implications for various indicators including welfare and employment. The study concludes in Chapter 8 by summarising the key findings and drawing attention to a set of policy implications emanating from the foregoing analyses.

2 Trade Liberalisation and Employment

A Review of Theory and Evidence

The nexus between trade liberalisation and employment has been a hotly debated and controversial issue in economic literature. The complexity of establishing the interface and linkage between trade reforms and employment is informed by the fact that there could be several channels through which trade liberalisation could influence the labour market and this could be direct, indirect and tangential. Two major streams of research can be discerned in the vast and rich literature that has emerged on this issue: the first one deals with the issue of job creation, destruction, relocation and associated adjustment costs originating from the opening up of the economy; the second one deals with the impact of trade liberalisation in terms of income, relative wage and distributional aspects. An attempt has been made here to review the existing literature both theoretical and empirical, focusing on both the aforesaid strands in the literature. It was felt that an overview of this literature will help better understand the cross-country experience and evidence, which would subsequently help in asking the right questions, analysing the available empirical evidence in Bangladesh context and eliciting useful policy conclusions for the present study.

2.1 Trade Liberalisation: Employment Creator or Destructor

A review of literature evince a general viewpoint that trade liberalisation is associated with both job destruction and job creation mainly because of the multidimensional adjustments that take place within the economy as a result of such reform. Employment effects could be positive or negative depending on country-specific factors such as the functioning of the labour and the product markets, as well as sector-specific factors within a single economy. When trade reforms take place and a country opens up, market forces bring structural changes in the economy by reallocating resources to more productive sectors. Efficiency gains from trade are derived only by paying the adjustment costs that could give rise to job losses. Particular sectors which are less competitive compared to global standards are more likely to suffer from competition under a trade reform policy. However, under standard assumptions about efficiency gains from trade, these losses are expected to be offset by gains in other sectors leading to positive overall employment effects, in terms of overall number of jobs, wages earned or a combination of both. However, other economists have argued that the net results of trade reforms in terms of employment will depend on many factors, and these would be different for different sectors of the economy.

The impact of trade liberalisation on employment is expected to vary between export-oriented and import-competing industries, and between labour-intensive and capital-intensive industries. Export-oriented sectors and import-competing sectors tend to have different productive and technological capacities. Trade liberalisation generally causes expansion of exporting sectors with higher productivity and increased demand for labour, while import-competing sectors, particularly which are not competitive enough and are generally protected, tend to suffer through reduced production and increased competition. This was likely to lead to a fall in employment in those sectors.

However, traditional trade models tend to assume that all workers were employed before trade liberalisation (full employment assumption) and that adjustment to the reform would take place without any time lag. Mussa (1978) showed for a traditional Heckscher-Ohlin (H-O) model with adjustment costs that an economy moves from an autarky equilibrium to a trade equilibrium if adjustment takes time and is costly. One of the possible outcomes is that the economy never manages to reach the trade equilibrium that would be optimal in the absence of adjustment costs.[1]

---

1. It is worth noting that the "sticky" production factor in this model is capital and not labour.

Given this assumption that laid-off workers would automatically move into new jobs, the issue of trade related unemployment did not arise. But this is not the case in reality. Furthermore, trade liberalisation also affects the quality of jobs. Since it was assumed that exporting and import-competing sectors employ different types of workers or employ them in different proportions, trade liberalisation in the long run affects the relative demand for different types of workers and this change in relative demand results in relative income changes (ILO, 2007). This conclusion has important policy implications since the inference is that the net job creation could be either positive or negative and that whatever the net result, this could have important implications for income distribution.

Some studies have argued that trade liberalisation creates employment opportunities mainly in the non-farm and non-tradable sectors, and this positively impacts on poverty reduction. When countries open up, they argue, the forces of globalisation helps the non-tradable sectors both from demand and supply sides by stimulating demand for the goods and services produced by these sectors and providing these with better access to inputs and machinery through import liberalisation (Osmani, 2005).

Some have argued that trade liberalisation may also create new import demand in the developed countries for products originated in developing countries which leads to employment creation in those countries (Lee and Ronald-Holst, 1995). Possibility of creation of employment opportunities through higher inflow of FDI, as a consequence of trade liberalisation, has also been highlighted in the literature (Arbache, 2001; Feenstra and Hanson, 1997; Zhu and Trefler, 2005). However, these studies also draw attention to the issue of wage inequality as a result of trade reform-driven FDI.

Indeed, the empirical literature on trade liberalisation-employment nexus evince that these studies have used quite a broad range of methodologies. Some cross-country macro level studies have examined the issue of whether or not the cross-country variations in employment performance could be explained by cross-country variations in the level of trade liberalisation. Some single-country macro level studies have also examined the impact of trade liberalisation measures on the employment performance of the economy under consideration, usually over time. As opposed to these macro level studies, there is a vast sphere of literature which has explored the impact of reforms at the sectoral or micro levels. However, concerns have been expressed as regards the appropriateness of the methodologies that have been employed in the empirical research in this context.

According to Arbache (2001), although trade liberalisation could result in higher output, lower domestic prices and increased labour demand and ensure overall economic improvement, it would actually depend on socio-economic-political context of a specific country. Fu and Balasubramanyam (2005) found a positive and significant impact of exports on employment. They showed that a 1 per cent increase in export volume raises employment by 0.17 per cent by using panel data set for TVEs (township and village enterprises) in 29 provinces in China over the time period 1987-1998. Manufacturing employment was found to have increased in the short-run significantly, at a time when trade liberalisation was being carried out in Mauritius (Milner and Wright, 1998), Costa Rica and Peru (Harrison and Revenga, 1995). According to a World Bank study on trade reforms carried out by Papageorgiou *et al.* (1990), for developing countries, in eight out of nine countries manufacturing employment was higher during and one year after the liberalisation period than it was before. The results of the Papageorgiou *et al.* (1990), however, were challenged later by Lee (2005) and on the question of methodology by Greenaway (1993) and Collier (1993). Greenaway (1993) pointed out that in Papageorgiou *et al.* (1990), either some kind of descriptive statistics (which is subject to subjective judgement) were used or they relied on ad hoc regression equations (predominantly bivariate) which are not very helpful. Besides, there was no effort to distinguish the effects of liberalisation from the effects of other policy changes. Collier (1993) also discussed the inherent drawbacks of this study and concluded "it lacks the analytical and quantitative rigour which might have made it definitive" (p.510). Matusz and Tarr (1999) pointed out that this World Bank study discussed only manufacturing employment and do not measure under-employment.

Harrison and Revenga (1995) considered total employment instead of only manufacturing employment; time period covered was 1984-1990. Besides, they did not undertake any regression analysis. The authors reported that total employment figure rose to 1136.2 thousand in 1990 from 932.6 thousand in 1984 for Uruguay. However, in a number of transitional economies such as Czechoslovakia, Poland and Romania employment fell during the transition period. Revenga (1995) found that in Chile, manufacturing employment has decreased significantly during and after trade liberalisation. Another study by Rama (1994) found a negative effect of trade liberalisation on employment in Uruguay in the late 1970s and early 1980s which contradicted with the findings of Harrison and Revenga (1995). Rama (1994) estimated the labour demand function for Uruguayan manufacturing sector, with the help of econometric exercise, using two different indicators for trade liberalisation (nominal and implicit protection rate). The time period for this study was 1978-1986 (a six-year panel). He found that reduction in tariff-inclusive price of imports by 1 per cent led to a drop in employment of 0.377 percentage points (in case of nominal protection) or 0.514 percentage points (in case of implicit protection).

Investigation by Greenaway *et al.* (1998) indicates that between 1979 and 1991, the UK industry had integrated into the international economy through FDI and trade that caused large scale job losses in UK manufacturing sector. According to Ernst (2005), impact of trade liberalisation on Argentina and Brazil was disappointing, however in the case of Mexico, growth in productivity and employment was found for the manufacturing sector for the period covering the second half of 1990s. Agenor and Aizenman (1996), however, pointed out that most of the empirical studies consider only the manufacturing sector and provide scant evidence on changes in employment in non-manufacturing production activities and changes in the aggregate unemployment rate. Again, Matusz and Tarr (1999) noted that in these country-specific studies, economies were also undergoing significant other reforms that went well beyond the ambit of narrow trade liberalisation.

According to a large number of studies, trade liberalisation had adverse impact on employment sector since developing countries have to go through structural adjustments and economic reforms to adjust with the trade liberalisation process (Arbache, 2001). Carneiro and Arbache (2003) found limited impact of trade liberalisation on macroeconomic variables and labour market indicators for Brazil. A World Bank study on globalisation takes a less sanguine view of the employment effects of trade liberalisation as distinct from an earlier study (Dollar and Collier, 2001; cited in Lee, 2005). It notes that a series of case studies on the effects of trade liberalisation showed considerable dispersion as regards the net impact on employment. Higher competition in goods markets, lower trade barriers and higher integration of goods markets across countries, more intense globalisation and greater outsourcing, are all leading to an adverse environment in terms of job destruction. According to some authors, when the environment becomes more volatile, existing labour market institutions may become dysfunctional and lead to substantially higher unemployment (Jansen and Turrini, 2004). According to Ljunqvist and Sargent (1998, 2005), uncertainties and turbulence do not lead to more volatility in the job markets but to stronger negative effects in case of job losses. Exchange rate was also identified as an important factor which created impact on employment especially in the industrialised countries (Hoekman and Winters, 2005; Klein *et al.*, 2003; Gourinchas, 1999).

Thus, with regard to possible impact of trade liberalisation on overall employment, the jury is still out and the results run both ways. Obviously, other factors tend to be at play and impact on the end-result in this context. However, there are a number of studies which tend to show that employment in export-oriented industries tend to increase following trade reforms but with regard to overall employment, the results appear to be ambiguous.

2.2 Impact of Trade Liberalisation on Wage Inequality

Though in many instances trade liberalisation was found to rise average income levels (i.e. wages), economists could not reach an agreement on the precise nature of this relationship.[2] The neoclassical approach to the issue tends to argue that cuts in real wages will increase employment implying that labour market reforms aimed at bringing more flexibility was important for improving the employment situation and generating growth. The neoclassical model usually applies a smooth labour supply function which increases with higher real wages while a smooth

2. See, for example, Baldwin, 2003; Dollar, 1992; Harrison, 1996; Rodriguez and Rodrik, 2001; Sachs and Warner, 1995; Dollar and Kraay, 2004; Loayza *et al.*, 2004 and Wacziarg and Welch, 2003.

labour demand function with inverse proportional relationship with real wages. Flexible real wages lead to market clearing—as in a normal goods market—and (near) full employment. However, these smooth functions are not based on neoclassical microeconomics. In complete harmony with the neoclassical model, a more differentiated supply function for labour is possible. When real wages become very low, any further reductions in real wages may increase the supply of labour.[3] On the other hand, increasing real wage rates do not increase labour supply, as leisure time becomes more attractive for members of the households. The neoclassical smooth demand function for labour is based on a macroeconomic production function that suggests that industry sectors with a representative firm use labour and one capital good for production. Such studies implicitly assume that there is no interaction between the sectors and the different markets. The inverse relationship between real wages and labour demand in the neoclassical model depends on the existence of only one capital good. Two or more capital goods destroy the relation between lower real wages and higher labour demand (Herr, 2002).

On the other hand, rejecting the neoclassical view the Keynesian approach states that there is a hierarchy of markets: the most important is the capital market in economic sense (including the banking system) while the labour market lies at the bottom of the hierarchy. This school of thought predicts that demand for labour depends on the volume of production and the technology. The wage level does not determine employment; rather, it is an important factor in determining the price level. In other words, reductions in wages do not lead to higher employment; they lead to deflation and to a destabilisation of the economy.

However, in empirical works on the trade-wage link, theoretical foundation predominantly lies mainly on the Theorem of Stolper and Samuelson.[4] Stolper and Samuelson (1941) concluded: "International trade necessarily lowers the real wage of the scarce factor expressed in terms of any good" (p.346). According to this theory, trade with the labour-abundant

3. This phenomenon could historically be noticed during the industrial revolution. In many transitional and developing countries, we can see a similar development. The phenomenon is also known in developed countries for the group of low paid workers.

4. See, for example, Deardorff and Hakura, 1994; Melchior, 2006.

country will lead to a reduction in the real wage of labour in the capital-abundant country. There is a corresponding magnification effect by which the relative change in factor prices is stronger than the relative change in goods prices (Jones, 1965).[5] It is common knowledge in the literature on the HOS (Heckscher-Ohlin-Samuelson) trade model that SST only applies when countries are diversified; e.g., produce both goods in the 2x2x2 model. Stolper and Samuelson (1941: 352) considered the case with complete specialisation and concluded that for a capital-abundant country, trade would reduce the real wage measured in terms of the exported good, but the impact expressed in terms of the imported good would be ambiguous. The focus has rather been on Factor Price Equalisation (FPE), which also rests on the assumption of diversified production. As noted by Bhagwati *et al.* (1998: 107), "Incomplete specialisation in equilibrium is considered the norm and complete specialisation ... is treated as an unlikely event."

According to HOS theorem, trade liberalisation has different impacts on developed and developing countries which predicts that increased trade between the two will lead to increasing wage inequality between less-skilled and more-skilled workers in developed countries but decreasing wage inequality between these two groups of workers in developing countries. Developed/industrialised countries are relatively well-endowed with skilled labour than the developing countries (Polaski, 2006) which could lead to increase in inequality.[6] In case of developing countries, it is important to understand which is the group for which group's wage is increasing: 'skilled labours' or 'unskilled labours'? It is expected that trade between industrialised and developing countries would lead to a decrease in the wages of low-skilled workers in industrialised economies and increase in wages in developing countries. After liberalisation had taken place, unskilled labour would suffer in developing countries and wage inequality would increase. Policies related to trade reforms could, therefore, have

---

5. In textbooks, it is the latter relationship that is presented as the Stolper-Samuelson Theorem (SST).

6. ILO's (2007) observation also confirmed that trade among industrial countries actually contribute to raise inequality as trade among similar countries could raise wage inequality within countries and also within sectors.

considerable employment impact in those countries. The issue was more prominent if the concerned economy was not characterised by (near) full employment before the reform had taken place or if domestic policies or labour market characteristics hampered the adjustment process for some reason or other. An ILO (2007) study makes the point that this situation was more likely in case of developing countries; under such circumstances, trade liberalisation would lead to increase in formal employment levels rather than resulting in changes in wages.

Significant developments have been made in recent years by Wood (1994, 1997) and Rodrik (1997) who have put forward alternate theoretical perspectives on trade-employment nexus. They argued that international trade may have adverse effects on wages and employment of unskilled workers by bringing about technological change. This was because such changes led to displacement of unskilled workers and/or by increasing the own-price employment elasticity of unskilled workers. This makes such workers more vulnerable to economic shocks. However, Sen (2003) argued that there is no theory that can unequivocally predict what would happen to wage inequality or more generally, to wages and employment in a developing country with increased international trade. Trade liberalisation helps to increase employment and wages in the short run in the exportable industries but expected to decrease in the long run since improvement of organisational capacity and learning by doing are involved at that stage. In case of import industries, employment increases both in short and long run whereas wages decrease.

In the case of developing countries, trade-wage inequality nexus has produced mixed results. Edwards (1997) did not find any significant effect of trade on income distribution as it was found that the link between income inequality and trade reform was largely country and situation specific. Mouelhi (2007) found a positive impact of trade liberalisation on unskilled labour demand in export-oriented manufacturing sector and a negative impact on domestically oriented firms in the case of Tunisian manufacturing industry (this is consistent with H-O theory). In case of East Asian countries, the gap in wages between skilled and unskilled workers had narrowed in the decade following trade liberalisation. For example, in Malaysia, differences between wages of skilled and unskilled workers decreased between 1973 and 1989; however, this proved to be ambiguous in case of Philippines (Wood, 1997).

Krugman (2000) and Arbache (2001) also tend to agree that increased volume of trade is an explanation of growing wage inequality. Latin American experience has come out as very interesting examples with marked differences in empirical findings across countries. On the one hand, researchers have found that trade openness affected income distribution positively (Dollar and Kraay, 2002) in Latin America (Behrman *et al.*, 2003). On the other hand, some of the other studies suggest that trade liberalisation has coincided with an increase in both income and wage inequality (Slaughter, 2000; Wood, 1997; Robbins, 1996; Sanchez-Paramo and Schady, 2003; Spilimbergo *et al.*, 1999; Barro, 2000). Trade, in particular, if combined with FDI, tended to increase volatility in the labour markets. More competitive product markets meant more competition which led to increased wages and costs that translated into larger declines in output and demands for all factors (Rodrik, 1997; Scheve and Slaughter, 2004). The different outcomes of Latin America and East Asia could be because Latin America had opened its market later compared to the East Asian countries (Wood, 1997). In case of China, at national level, inequality has increased with the opening up of the economy (Gini coefficient rose to .0452 in 1995 from .0382 in 1988 which was 0.288 in 1981); at the same time, Chinese cities that have had a greater increase in the trade-to-GDP ratio have witnessed a reduction, rather an increase, in income inequality (Wei and Wu, 2001).

Currie and Harrison (1997) argued that the wage impact of liberalisation depended crucially on the nature of product market competition. If increased product market competition reduces the relative price of low skill-intensive products, trade could generate perverse wage-inequality effects. They argued that this competitive effect of liberalisation was found in Morocco. The explanation for this phenomenon was that, jobs which were low skill-intensive in the United States were relatively skill-intensive in a country like Mexico. The literature showed that wage inequality has increased in Mexico and Morocco because protection was higher in the low skill-intensive sectors which actually led to shrinkage of these sectors (Slaughter, 2000).

Intra-industry trade, therefore, would not create the changes in relative demand for different production factors predicted by the Stolper-Samuelson Theorem. But Manasse and Turrini (2001) found a different picture that indicated that intra-industry trade increased wage inequality within countries and also within sectors. Interestingly, it has also been argued that for this wage inequality, not trade but technological progress was responsible to a greater extent (Duranton, 1999; Berman *et al.*, 1994; Lawrence and Slaughter, 1993) though Wood (1994) attributes 70 per cent of the causation to trade which increases the demand for skills in various regions (Zhu and Trefler, 2005). In the US, one-third of the net increase in the skilled/unskilled wage ratio from 1973-1993 was attributable to trade (Cline, 1997). Attanasio *et al.* (2003) found a link between trade, skill-biased technological change and increase in wage inequality and argued that drastic tariff reduction affects wage distribution. It actually negatively affects those industrial sectors that have lower wages and a higher fraction of unskilled workers. When trade liberalisation occurs, labour force shifted towards the informal economy that typically paid lower wages and offered no benefits. With respect to industry wages, they found that wage premiums decreased more in sectors that experienced relatively larger cuts in tariffs.

The above review of the literature on impact of trade reforms on employment, income, wage and inequality tends to indicate that there is no overriding single trend of results. The impact was determined by depth and coverage of reforms, technological change, nature of product competition and several other factors. Evidence suggests that the nature of relationship is largely and significantly country-specific. This is important to note as this would lead to two important conclusions. Firstly, there is a need to study and establish the nature of these relationships based on specific country experience; and secondly, policy conclusions and design of policy responses needs to be made on concrete country-specific experiences.

From the above literature review, it is clear that the results are far less than conclusive as regards the trade liberalisation-employment-wage nexus. Findings vary not only from country to country which is quite natural because of the differences in factor endowments in different countries. Indeed, evidence sometimes tends to differ for a particular country as well. The point that needs to be reemphasised is that findings of the empirical studies (both country-specific case studies and cross-country analyses) may differ significantly due to methodological differences (e.g., use of different theoretical framework, application of statistical techniques, model specifications etc.), different time coverage, use of diverse range of indicators for trade liberalisation, sectoral coverage (i.e., manufacturing or total employment, macro/sectoral/micro data) etc.

2.3 Impact of Trade Liberalisation on Bangladesh Economy: Evidence from Studies

Several studies have looked at the impact of trade reforms in Bangladesh, from a number of perspectives. Some of these were descriptive while others were more rigorous and analytical. Some of these studies have also used CGE models to measure the likely outcome of such reforms.

Mujeri and Khondker (2002) looked at the poverty implications of trade liberalisation in Bangladesh applying CGE model. They found positive but relatively small gains from trade liberalisation from the perspective of alleviating the poverty situation of the country. They observed that due to structural bottlenecks and other constraints, the gains from trade liberalisation accrued more to the relatively well-off households compared to poor households. According to Ahmed and Sattar (2003), trade openness had positive impact on poverty reduction and overall growth in Bangladesh. They have found that poverty has reduced significantly during the period 1976-1990. However, others have pointed out that during the same period income inequality in Bangladesh has also increased (Osmani, 2005). Nath and Mamun (2005) have also found that trade plays an important role in growth but it also increases income inequality. However, Durevall and Munshi (2006) found no relation between trade liberalisation and changes in wage inequality which runs contrary to traditional HOS theory. In case of skilled and unskilled labour, real wage had increased due to skill-neutral technical change. However, the authors concede that the result may be different for the informal sector.

In Bangladesh, the contribution of international trade to total employment growth has been positive during the year 1990 due to significant increase in both labour-intensive manufacturing exports and labour-intensive imports. Sen (2003) found that increased integration with the world economy has led to increase in manufacturing employment and adoption of more labour-intensive techniques within the same industry in Bangladesh.

Hoque (2006) observed significant sector-wise variation in terms of impact of trade liberalisation in Bangladesh. Output and employment in export-oriented industries have experienced positive growth. However, industries which received low levels of protection posted higher growth rate although highly protected industries have suffered significantly when the domestic market opened up.

According to Khan (2000), trade reforms have not encouraged development of an autonomous domestic industrial capability in Bangladesh, indicating that creation of employment opportunities have been limited. Though export sectors such as readymade garments (RMG) benefited significantly, employment in import-competing sectors like textiles, chemicals and metals have been adversely affected during this period. Muqtada (2003) also observed that employment increased in the export-oriented sectors but found that except for the RMG sector, employment situation in other manufacturing sectors did not improve as was expected. Muqtada cited import competition, public sector downsizing and restructuring of private sectors as factors responsible for this. According to Rashid (2000), trade liberalisation negatively affected small and cottage industries of Bangladesh. However, others have argued that trade liberalisation was unable to have positive impact on most of the industrial sub-sectors because Bangladesh's marketing, technical and entrepreneurial skills were not developed enough (Khan, 2000). Rashid (2000) put forward the suggestion that Bangladesh should emphasise on skill development of its labour force and should introduce new and more efficient technology for increasing labour productivity.

A review of relevant literature suggests that studies have mostly focused on aggregate level dynamics in the labour market and have tried to relate and explain this from the perspective of overall change in the trade regime. Some studies have looked at changes in the labour market at some levels of disaggregation but rigorous analysis to capture the relationship between dynamics of trade liberalisation and labour market movements is largely absent.

3 Trade Liberalisation in Bangladesh

Phases, Pace and Sequence

Trade liberalisation policies adopted by Bangladesh since early 80s have been discussed in this chapter to set the tone for subsequent analysis with regard to the impact of those reforms on the labour market dynamics in Bangladesh. A distinctive feature of this section is that the entire trade liberalisation period, starting from 1980 to the present time, has been analysed based on a range of indicators to understand the major trends, pace and sequencing of trade reform policies pursued by successive Bangladesh governments over this period. An attempt has been made to identify distinct phases of the trade reforms and document the major reform policies pursued during each of the phases.

3.1 Brief History of Trade Liberalisation in Bangladesh

The decade of 1980s, and more particularly the 1990s, has experienced an unprecedented, wide ranging and speedy move towards openness of economies driven by trade reforms. The majority of the developing countries have moved away from import-substituting, public sector-led, inward-looking development strategies to an export-oriented, private sector-led and outward-looking one. In majority of the cases, this move has been supported as part of package of conditionalities under the World Bank-IMF aid programmes. It would be pertinent to recall in this context that ever since the 1960s, a view which gathered momentum in the World Bank was that a more liberal trade regime would be conducive to better economic performance (Edwards, 1997). This particular view was strongly supported by Balassa (1971) and others who propagated the hypothesis that protectionism seriously hampered economic performance and growth prospects of developing countries. In the 1980s, the World Bank and the IMF initiated a series of loan packages for policy support, better known as the Structural Adjustment Facility (SAF) and Enhanced Structural Adjustment Facility (ESAF) that were aimed at underwriting radical reform measures in developing and transitional economies. One of the most important policy prescriptions championed through these policy-based lending was the liberalisation of domestic trade regime and through it, creation of an enabling environment for development of domestic entrepreneurship and attracting foreign investment flow. A majority of developing countries, including Bangladesh, have implemented structural adjustment reforms in the latter half of the 1980s, by way of encouraging reduction and dismantling of both the quantitative and qualitative restrictions on trade. Although the developing countries, and particularly LDCs such as Bangladesh, were exempted from taking any tariff reduction commitment as part of the GATT/WTO negotiations, reforms under the aforesaid SAP conditionalities prepared the ground for significant reduction of both tariff and non-tariff barriers.

3.2 Phases and Pace of Trade Liberalisation

Bangladesh's trade regime has passed through significant changes since the 1970s. In years immediately following independence, Bangladesh has pursued a highly restrictive trade policy with high effective rates of protection and import substituting industrial policy being its major hallmarks. Some changes in the policy started to emerge following major political changes in mid-1970s. Indeed, since then trade liberalisation has become a constant feature of Bangladesh's policies. The speed and structure of reforms have, however, not been uniform. Several distinctive phases could be identified in this connection though not all authors agree with the time lines for the respective phases. In an earlier study, Hussain and Rahman (1995) stated that the agenda for trade liberalisation moved forward in three phases: FY 1982-1986 (first phase), FY 1987-1991 (second phase) and FY 1992 onwards (third phase). Raihan (2007) maintained that in Bangladesh, three broad regimes of trade policy reforms can be defined: from 1972 to 1980—the regime of restricted trade; from 1980 to 1991—the regime of moderate trade liberalisation; and from 1991 onwards—the regime of rapid trade liberalisation. Broadly agreeing with the views expressed in the above studies, it is argued here that in order to relate the labour market dynamics with reform programmes, it would be appropriate and useful to divide the post-independence period into four phases based on the basis of speed of the trade reforms and major indicators of reforms that were implemented over these years.

As was mentioned above, trade reforms in Bangladesh were taken in earnest in early 1980s. Accordingly, the pre-1980 period could be termed as pre-reform phase. Post-1980 period (till now) could be divided into three phases. The first reform phase identified in this chapter is for the period between 1981 and 1990 which was a regime of moderate trade liberalisation. The restrictions on trade started to be relaxed in this decade; new industrial policies were put in place to stimulate private investment. Tariff rates, traditionally very high, started to come down and some of the import restrictions started to be lifted. A new industrial policy was formulated in 1982 (NIP-82) and a revised policy was designed in 1986 (RIP-1986). Investment ceiling was raised, sectors were opened up for investment, private sector was encouraged to invest, restrictions on FDI were lifted and there was a shift from positive list to a negative list.[1] Bangladesh also contracted a three-year SAF loan with the IMF (1986-1989) in this period.

During the second phase i.e., from 1991 to 1995, the speed of reforms significantly intensified. After a year of hiatus, the country entered into an ESAF programme (1990-1993). This period of trade liberalisation in Bangladesh was characterised by a host of trade reform measures which included significant decline in quantitative restrictions, tariffication of quantitative restrictions and rationalisation of import tariff, opening up of trade in many restricted items and liberalisation of the foreign exchange regime (Razzaque *et al.*, 2003). As a result, there was a significant reduction in the anti-export bias in the trade policies and also in the effective rate protection (ERP) of domestic industries. There was also a significant reduction in tariff dispersion (the so-called water in the tariff). World Bank (1998) estimated that in this period, after Columbia, Bangladesh was the fastest liberalising country in the developing world.

1. A Positive List meant that sectors identified in the list would be open to private FDI. A Negative List indicates that all sectors barring those in the Negative List will be allowed for private FDI.

The third phase could be said to have initiated in 1996, when the reforms were mainly locked in but the pace of reforms itself slowed down following the very fast pace of reforms in the second phase. Some of the trade liberalising measures continued in this phase including further liberalisation of import duties, dismantling of non-tariff barriers and quotas and reductions in the number of income tax slabs.

As was noted in Chapter 1, there is a serious debate with regard to appropriateness of indicators to measure the extent and speed of trade liberalisation. Indeed, a large body of literature exists on various indicators of trade reforms with their relative merits and demerits which include tariff, trade, exchange rate and many other policy related variables. Not surprisingly, these indicators come up with diverse results because of their methodological and conceptual differences and create confusions among the researchers. Apart from methodological and conceptual issues, lack of transparency and harmonisation in the tariff regime of various countries (for example, use of various country specific para- and non-tariff measures, preferential tariff rates etc.) have contributed further to this paradox. Besides, in many cases the indicators which would be conceptually the most appropriate and suitable, cannot be constructed or estimated due to data unavailability. This study, however, does not attempt to examine the effectiveness and appropriateness of a particular indicator of trade reforms in the context of Bangladesh; rather it has used several indicators of trade liberalisation to understand the pace and depth of the reforms and justify the various phases identified above. These are:

- Tariff reform indicators
- Trade related indicators and
- Other policy related indicators.

In the next section, various indicators during these phases of trade liberalisation have been discussed and their pace and sequence examined.[2] Table 3.1 shows the phase-wise averages (which one is applicable) of some key trade indicators[3], the trade also shows annual growth rate of some of the indicators. It may be observed from the table that in the second phase,

2. Please see Annex Tables A-3.1, A-3.2 and A-3.3 for further details.

3. For the definition of indicators please see Annex A-3.1.

the tariff related indicators have come down significantly while trade has experienced a significant growth.

Table 3.1

Phase-wise Per Annum Growth and Averages of Major Trade Liberalisation Indicators

	Pre-Reform Before 1980s	*Phase One 1981-1990*	*Phase Two 1991-1995*	*Phase Three 1996 and onwards*
Per annum average (%)				
Simple average tariff rate	98 (1973-1980)	100.25	51.08	17.61 (1996-2007)
Import weighted average tariff rate	Na	Na	26.96	12.69 (1996-2007)
Implicit nominal tariff rate	18 (1977-1980)	15.92	17.37	13.9 (1996-2004)
Trade-GDP ratio	24.61 (1980)	22.03	19.8	31.47 (1996-2004)
Import-penetration ratio	18.69 (1980)	16.42	12.61	18.85 (1996-2004)
Export-orientation ratio	5.92 (1980)	5.61	7.2	12.62 (1996-2004)
Trade tax % of GDP	3.23 (1980)	2.72	2.29	2.07 (1996-2005)
Terms of trade	Na	92.78 (1986-1990)	98.62	98.22 (1996-2002)
Per annum growth rate (%)				
Total trade	13.41 (1974-1980)	5.00	12.23	8.06 (1996-2006)
Exports	10.04 (1973-80)	7.93	15.12	9.47 (1996-2006)
Implicit nominal tariff rate	-4.6 (1977-1980)	-0.062	-1.39	-0.70 (1996-2004)

Note: % growth per annum has been estimated as: [exp {ln (1+r/100)/n}-1]*100. Since comparable GDP data for Bangladesh before 1980s is not available, for pre-reform period ratios involving GDP data, figures of 1980 has been taken as the period average.

Source: Authors' estimation based on data collected from the various issues of BBS, BB, BER, EPB, BTC, NBR.

Table 3.1 presents salient features of tariff reform indicators in the four identified phases. As would be seen from the table, in terms of pace of reduction of the tariff, phase two stands out as the most liberalising phase in post-independence Bangladesh.

3.2.1. Tariff Reform Indicators

Significant reforms in tariff structure were initiated in the early 1990s. Tariff bands were reduced to 5 (in 2004-05) from 18 (in 1991-92). At the same time, unweighted tariff rate (per cent) and maximum rate (per cent) have also declined significantly. Most notable decline was observed in the highest tariff rate which has come down from 350 per cent to 25 per cent between 1991/92 and 2004/05 (Table 3.2).

Table 3.2

Tariff Structure of Bangladesh

Year	*Number of Tariff Bands*	*Maximum Rate (%)*	*Unweighted Tariff Rate (%)*
1991-92	18	350	70
1992-93	15	300	47.4
1993-94	12	300	36
1994-95	6	60	25.9
1995-96	7	50	22.3
1996-97	7	45	21.5
1997-98	7	42.5	20.7
1998-99	7	40	20.3
1999-2000	7	37.5	19.5
2000-01	5	37.5	18.6
2001-02	5	37.5	17.1
2002-03	5	32.5	16.5
2003-04	5	30	15.6
2004-05	5	25	13.5
2005-06	4	25	13.41

Source: Authors' estimation based on data collected from BTC and the various issues of BER.

Besides the indicators presented in Table 3.2 (tariff bands, unweighted tariff rate), several other indicators for tariff reforms have also been discussed in the following sub-sections.

a) MFN (Simple) Average Tariff Rate: The simple average tariff rate is the unweighted average MFN applied duty rate averaged over all the tariff lines for goods imported into the country. It is considered to be a good indicator of trade liberalisation if and where variations in the rates are not significant. However, when variations are substantial (high tariff rates for some goods and low for some other), a problem of underestimation (for high tariff) and overestimation (for low tariff) could arise. Figure 3.1 shows changes in simple average tariff over time for Bangladesh during the post-independence period.

Figure 3.1

Trend in Simple Average Tariff Rates

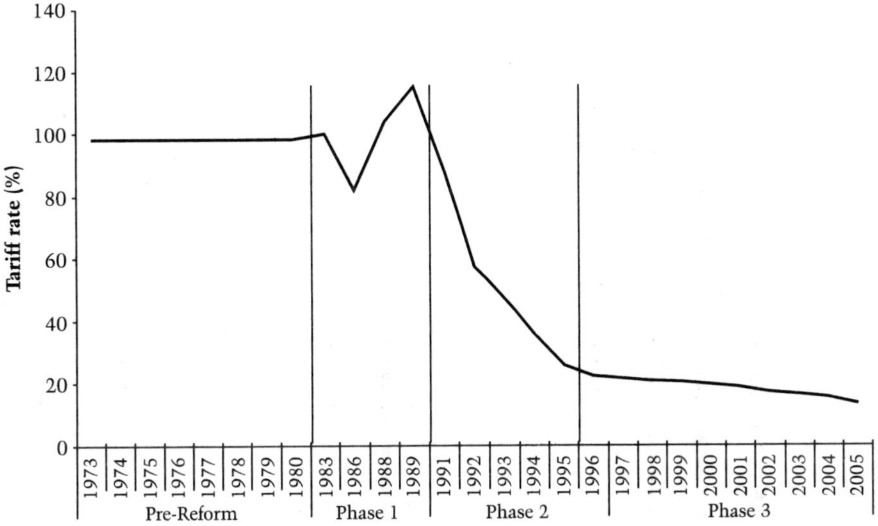

Source: Authors' estimation based on data collected from the various issues of BER and NBR.

The figure shows that before 1980s, the average tariff rate was about 100 per cent indicating a highly protectionist regime. Average tariffs started to fall from early 1980s although it spiked towards the end of the decade. During the second phase (from 1991 to 1995), the rate came down quite significantly, from around 88.6 per cent to about 22.3 per cent towards the end of the phase. The pace of decline slowed down in the third phase although the reductions have continued.

b) Import Category-wise Extent and Pace of Tariff Reform: Information in Annex Table A-3.4, Figures 3.2 and 3.3 suggest that, as a consequence of tariff reforms, both average unweighted and weighted tariffs have declined in Bangladesh quite significantly across all commodity categories (primary goods, intermediate products and capital goods) of imports. These duties have fallen most notably during the second phase of reforms. For example, for primary goods the unweighted import duties have fallen from 55.2 per cent to 17.8 per cent; and for final goods the corresponding figures were 80.2 per cent and 18.1 per cent in between 1991-92 to 2005-06. The fall was most rapid for capital goods during the same period, unweighted tariff rates of which came down from 45 per cent to 7.45 per cent. Similar trend is also evident for weighted average tariffs.

Figure 3.2

Import Category-wise Extent and Pace of Tariff Reform: 1991-2006

(a) Unweighted Average

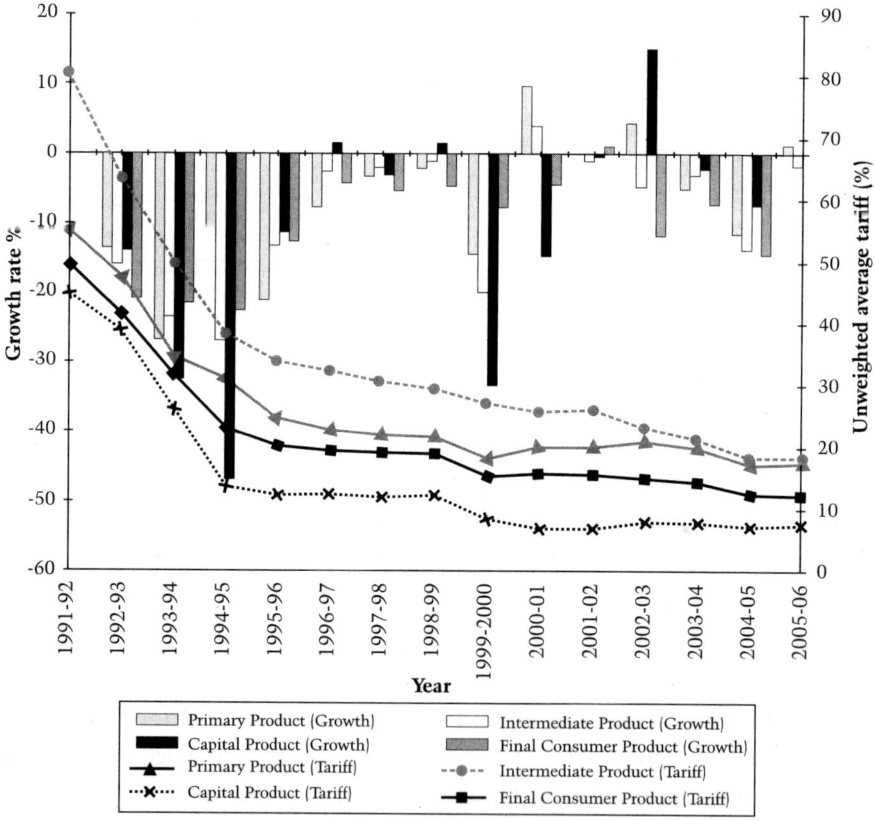

Source: Various Issues of *Bangladesh Economic Review*.

(b) Weighted Average

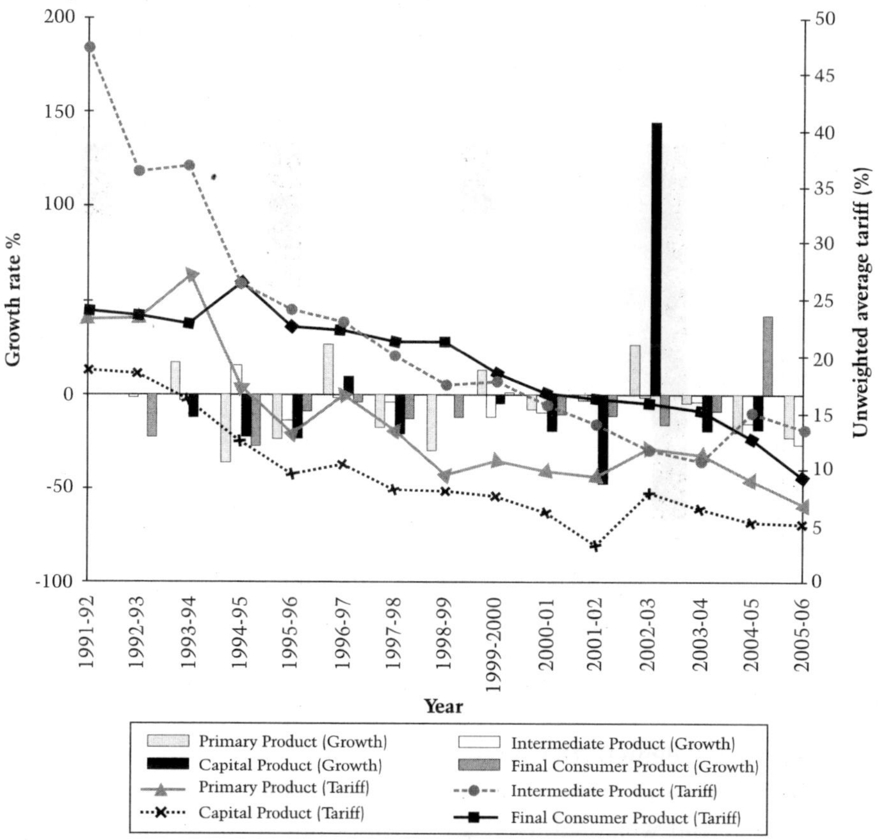

Source: Various Issues of *Bangladesh Economic Review*.

A review of the Figure 3.2(b) shows that the weighted average tariff rates for final goods came down from 47.3 per cent to 13.4 per cent between 1991/92 and 2005/06; for intermediate goods this was 24.1 per cent to 9.3 per cent and for capital goods this was 18.7 per cent to 5.2 per cent. One could conclude from this that Bangladesh followed the classical sequence of higher protection rate (higher tariffs) for final goods (13.4 per cent in 2005-06), relatively moderate tariff for intermediates (9.3 per cent) and relatively lower tariffs for capital goods (5.2 per cent).

Figure 3.3

Phase-wise Annual Compound Growth Rate (%) for Reduction of Average Tariff by Import Category

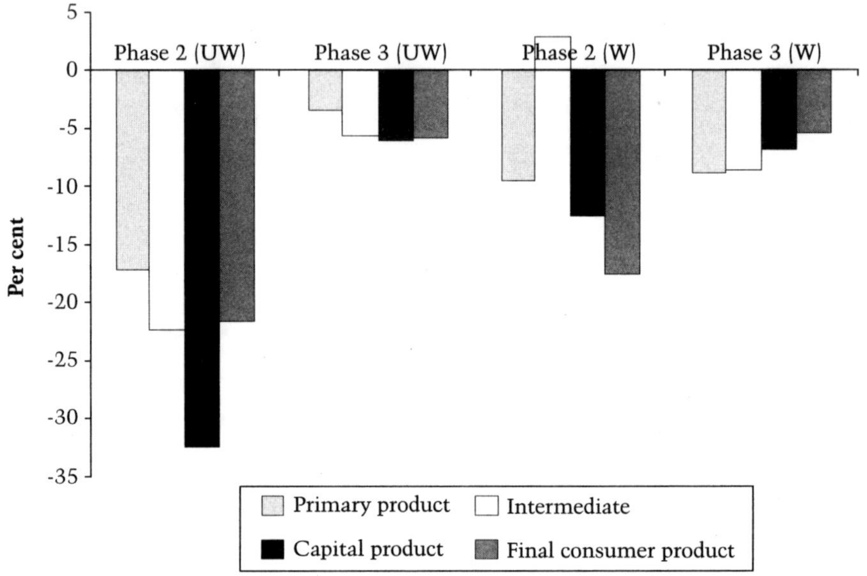

Source: Authors' estimation based on data collected from the various issues of BER.

Figure 3.3 depicts the phase-wise compound annual growth rate (CAGR) for various product categories for reduction of both simple (unweighted) and weighted averages of MFN import duty consequent to tariff reforms. With the exception of intermediate products in terms of weighted average during second phase, all other cases demonstrated higher rate of decline in tariffs in second phase as compared to third phase. If the unweighted average duty is considered, import duty on capital goods lowered with greater pace followed by intermediate and final products in both phases. It is to be, however, noted that higher cut in unweighted average tariff is observed in case of phase two. On the other hand, if the weighted average is taken into consideration, during the second phase protection declined most for final consumer goods followed by capital products; and surprisingly intermediate goods posted a positive growth. However, in the third phase the order of categories in terms of reduction rate for weighted average import duty has reversed; i.e., it was higher for intermediate goods (-8.6 per cent) compared to capital goods (-6.9 per cent) and final goods (-5.5 per cent).

Thus, it is seen that when the entire period of trade liberalisation in Bangladesh is considered, the country seems to have followed the classical sequence of relatively lower protection for capital goods and intermediate products and relatively higher protection for final goods. However, when each of the phases is considered separately, no particular such sequence can be discovered. Although this is more true for weighted average tariffs; for unweighted average tariffs, the classical sequence is visible even in case of individual phases of liberalisation.

c) Import Weighted Average Tariff Rate: The import weighted average tariff rate is a better estimate of average tariff compared to simple (unweighted) average tariff, as it better reflects the relative importance of various tariff lines and is calculated by weighting the tariff lines with their respective share in total imports. However, there is still a downward bias (even if there is no incidence of prohibitive tariff which will have a zero weight), because products with higher tariff rates tend to be imported in lower-than-otherwise quantities.

Figure 3.4

Trend in Import Weighted Average Tariff Rate during 1990s

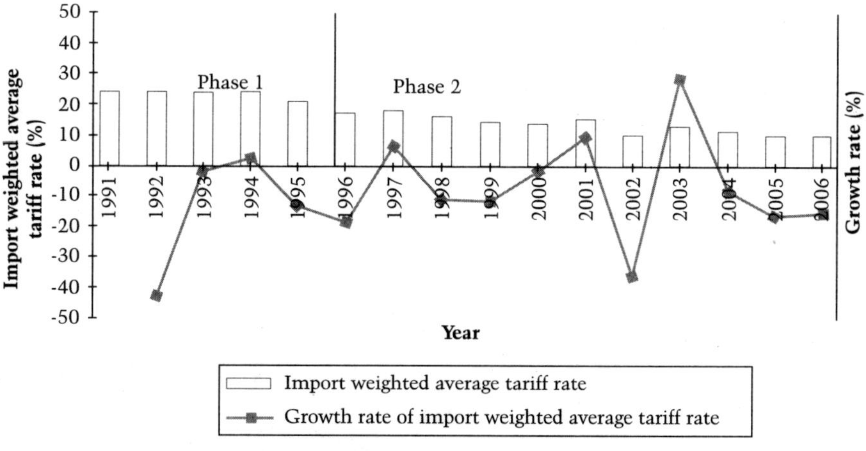

Source: Authors' estimation based on data collected from the various issues of BER.

Figure 3.4 shows the declining trend in import weighted average tariff rates in Bangladesh. This has declined from 42.1 per cent in 1991 to 6.98 per cent in 2007. During the second phase of trade liberalisation, the rate has declined quite sharply at an average rate of (-) 13.06 per cent. The declining trends have continued in the third phase as well, albeit at relatively slower pace (an average growth of (-) 7.14 per cent per annum). However, it is to be also noted that this overall declining trend in the growth rate have showed erratic movement in both phases, even registering positive growth in particular years.

d) Implicit Nominal Tariff Rate: It is the ratio of revenue (collected from import duties) to the CIF value of imports and a widely used measures of nominal protection (Greenaway and Milner, 1993). It is often argued that this is a more reliable measure of trade restriction as it takes into account the imports at duty free and preferential rate and considers the actual collection of custom revenues (Milner and Morrissey, 1999). It is a weighted average of all the rates that apply, with the weights being the share of imports from each source. However, this measure is sensitive to aggregation, mis-invoicing and custom related corruption.

Figure 3.5

Trend of Import Duty and Implicit Nominal Tariff Rate

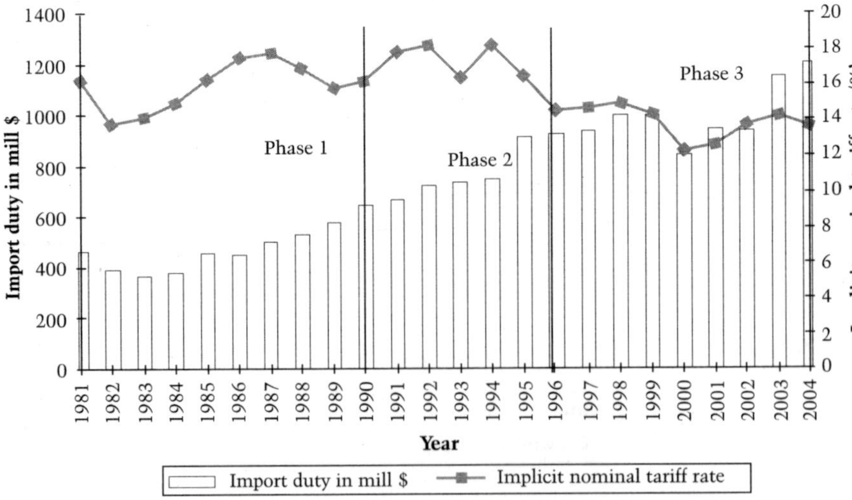

Source: Authors' estimation based on data collected from the various issues of BER and NBR.

Figure 3.5 shows the trend of collected revenue from import duty and implicit nominal tariff rate during different phases of trade liberalisation. Despite the fact of reduction in import duties, the revenue collection from import indicates, in general, an upward trend because of the robust growth of imports throughout this period. On the other hand, as revealed in earlier tariff reform indicators, this indicator has also showed, in general, a negative growth over the various phases of liberalisation period. The implicit nominal tariff rate was as high as 19 per cent in 1977. Throughout the first phase, this was hovering around 14-17 per cent (with an average of 15.82 per cent) while the second phase experienced an upward trend with an average of 17.4 per cent. However, during the third phase (1996-2004), the implicit nominal tariff rate experienced a declining trend within the range of 12.3 to 14.6 per cent and the average for this period again dropped below 14 per cent.

In early 1980s, the range for implicit nominal tariff rate is found to be between 13 to 16 per cent. Subsequently, it fluctuated between 13 to 16 per cent (in the decade of 1980s). During the second phase of trade liberalisation, upward trend is observed in 1991 and 1992 with a decline to 14.6 per cent in 1996; afterwards it remained within the range of 13 to 14.6 per cent. In late 2005, it has decreased to 13.2 per cent.

e) Quantitative Restrictions: Quantitative restrictions are important for an understanding about the extent and impact of non-tariff barrier. In Bangladesh, quantitative restrictions have been largely eliminated as a result of the trade reforms. Number of QRs has declined from 433 in 1989 to 63 in 2003-2006. In 1980s, around four-fifths of the QRs were imposed for trade reasons. From 1990s QRs for both trade and non-trade reasons have declined; for example in 2003-2006, out of a total of 63 QRs only 23 (36 per cent) were for trade related reason compared to 74 per cent in 1990/1991 (Table 3.3).

Figure 3.6 illustrates that from phase one to phase two, trade related restrictions as per cent of total HS 4-digit import lines has decreased from 40 per cent in 1986-87 to less than 2 per cent in 2003-2006. A rapid decline is observed from end of phase one of trade liberalisation, subsequent to which the trend has stabilised.

Table 3.3

Trend in Quantitative Restriction (At 4-Digit HS Code)

Year	*Total QRs in Place*	*% of QRs in Total No. of Commodities*	*Trade Reasons*			*Non-Trade Reasons*
			Banned	*Restricted*	*Mixed*	
1985-86	478	38.54	275	138	16	49
1986-87	550	44.35	252	151	86	61
1987-88	529	42.67	257	133	79	60
1988-89	433	34.91	165	89	101	78
1989-90	315	25.4	135	66	52	62
1990-91	239	19.27	93	47	39	60
1991-92	193	15.56	78	34	25	56
1992-93	93	7.5	13	12	14	54
1993-94	109	8.79	7	19	14	69
1994-95	114	9.19	5	6	12	92
1995-1997	120	9.67	5	6	17	92
1997-2002	124	10	5	6	17	96
2003-2006	63	5.08	5	8	10	40

Note: A total of 1240 4-digit tariff heading exists under the HS.

Source: Authors' estimation based on data collected from the various issues of BER.

Figure 3.6

Trade Related Restrictions as Proportions of Total 4-Digit HS Codes

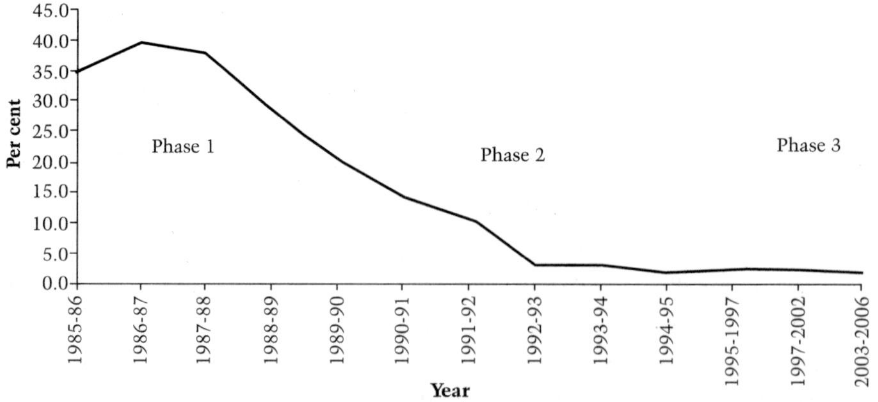

Source: Authors' estimation based on data collected from the various issues of BER.

3.2.2. Trade Related Indicators

A review of trade performance related indicators such as growth of exports, degree of openness, import penetration ratio, export-orientation ratio and trade tax as a proportion of GDP would allow to understand the implications of trade liberalisation for key trade indicators. However, it should be stated upfront that at this stage it will not be justified to draw any casual link between the two sets of variables.

a) Growth Rate of Exports: Export-oriented sector is generally considered to be one of the major beneficiaries of trade liberalisation. Throughout the liberation period, export has experienced robust growth. However, as is known, most of this growth in Bangladesh is accounted for by the RMG sector mainly by global market opportunities. Import of raw materials at zero-tariff (through the bonded warehouse facilities) and zero-duties on capital machineries for export-oriented industries have helped this sector to achieve phenomenal growth rates.

Figure 3.7

Trend in Growth Rate of Exports

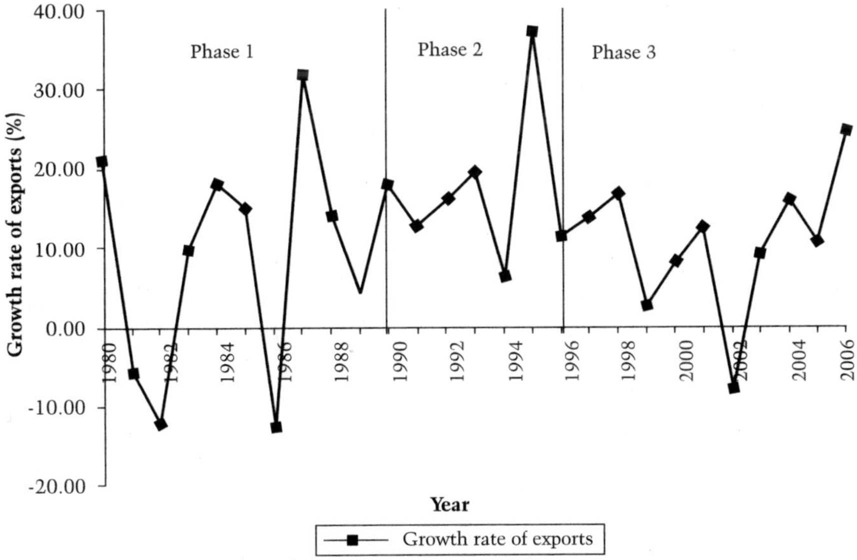

Source: Authors' estimation based on data collected from the various issues of BER.

As the trends evidenced by Figure 3.7 indicate, it is difficult to discern a clear-cut trend in terms of export growth performance for the various phases of trade liberalisation. However, on average, the percentage growth per annum of export was higher in the second phase of trade liberalisation; subsequently to this, the growth rate has declined in phase three (Table 3.1).

b) Degree of Openness: This indicator has been most extensively used in empirical research as an indicator of trade liberalisation. However, it has a number of inherent problems. Trade-GDP ratio is often influenced by structural factors of a country rather than differences in trade restrictions. As Rodriguez and Rodrik (2001) have argued that cross-country variations in the trade-GDP ratio may be primarily influenced by structural factors, such as geographical location rather than differences in trade restrictions.[4] In Bangladesh, during mid/late 80s there was a rapid expansion of "cut, make and trim" (CMT) assembly (by RMG) alongwith rapid expansion of the knitwear industry. Other notable exports are shrimps, leather and jute goods and raw jute. The export sector benefited from the removal of anti-export bias (through BWF facilities, zero tariff on imports of raw materials machineries, duty drawback facilities). Imports were encouraged through lower tariffs, removal of non-tariff barriers and by facilitating import of raw materials for export-oriented industries through B/B L/C facilities (Figure 3.8).

The rapid expansion of trade, both export and import component, has caused a rapid increase in trade-GDP ratio during the various phases of trade liberalisation.

Trade-GDP ratio was found to be within the range of 17 to 24 per cent in 1980s. However, it started to steadily rise since 1990. The rise is more visible in the third phase, since 1996. It is to be noted here that growth rate of export outpaced that of imports during this time.

c) Import Penetration Ratio: The import penetration ratio is the share of total imports in GDP which is expected to increase following trade

---

4. Raihan, S (2007). *Dynamics of Trade Liberalisation in Bangladesh: Analyses of Policies and Practices.*

Figure 3.8

Trend in Degree of Openness (Trade-GDP Ratio, in Per cent)

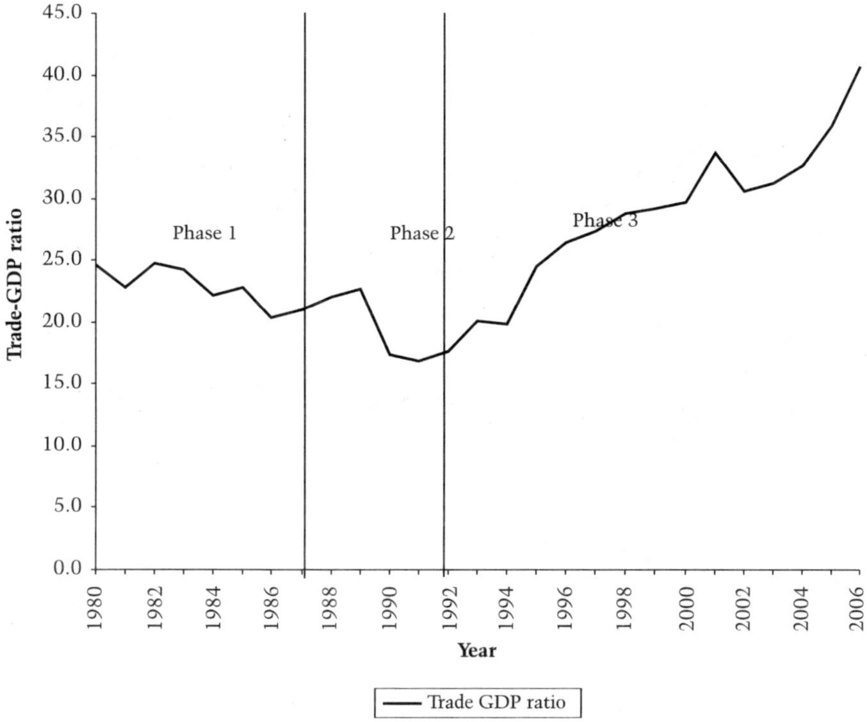

Source: Authors' estimation based on CPD database and data collected from the various issues of BER.

liberalisation. The advantage of including this ratio is that it captures the impact of both tariff and non-tariff barriers. Figure 3.9 displays the import penetration ratio for total imports and consumer goods. During 1980s, the range was between 12 to 17 per cent; however, in the early 1990s (second phase), the ratio had decreased (mainly in 1991 and 1992). On the other hand, in the third phase it showed an increasing trend.

In case of import penetration ratio for consumer goods was quite stable, ranging between 2 to 5 per cent, throughout the period 1980 to 2004.

Figure 3.9

Trend in Import Penetration Ratio for Total Imports and for Consumer Goods (in Per cent)

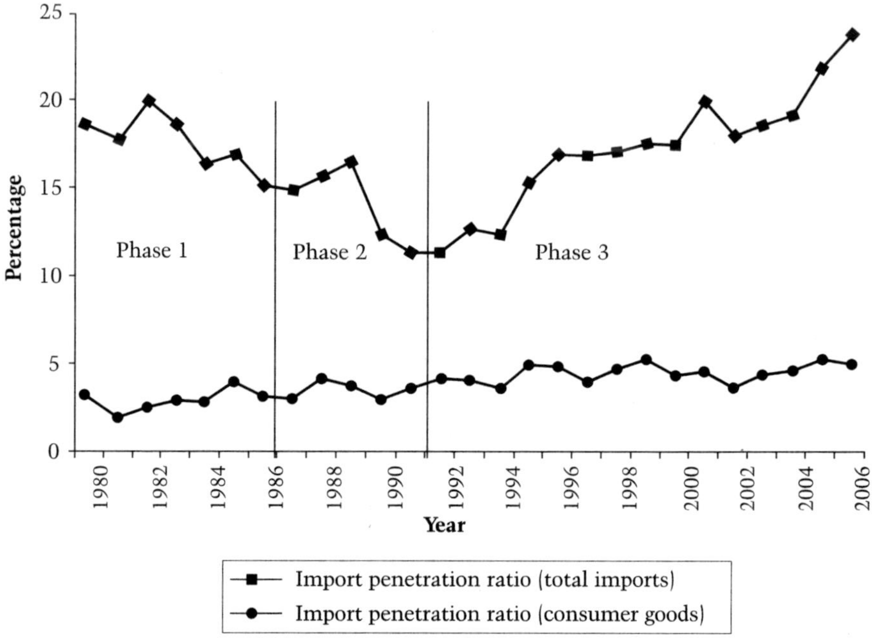

Source: Authors' estimation based on data collected from the various issues of BBS and WB.

d) Export Orientation Ratio (EOR): EOR is the ratio of total exports to GDP. During 1980s, the range was between 4 to 6 per cent. In 1995 and 1996, this was about 9.5 per cent of the GDP. By 2006 it has risen to 16.95 per cent. As was mentioned earlier, exports are mostly accounted for by export of RMG (76 per cent of total export). Besides reaping the benefits of trade reforms, the sector has also gained from MFA quotas in the US (providing market security) and from preferential treatment in the EU (thanks to zero-tariff market access under EU's GSP scheme). Figure 3.10 illustrates the trend in export orientation ratio of manufactured goods and export penetration ratio.

e) Trade Tax as a Per cent of GDP: One of the strong arguments against trade liberalisation originated from the strong dependence that the developing countries had on trade related taxes. This was the most

Figure 3.10

Trend in Export Orientation Ratio for Total Exports and Manufactured Exports

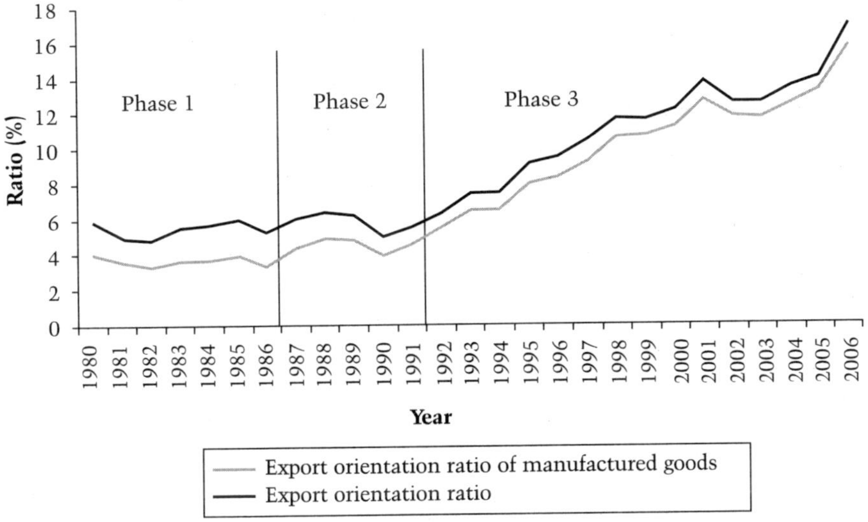

Source: Authors' estimation based on data collected from the various issues of BER and WB.

important source of revenue generation. The likely revenue loss from liberalising the trade regime was and continues to remain a major concern for the developing countries. Figure 3.11 represents trade tax as per cent of GDP, total tax revenue as per cent of GDP and the revenue generated from customs duties and other duties. Since along with direct tariff (i.e., custom duties), other duties (VAT, supplementary duty, IDSC etc.) constitute a large portion of the trade revenue, revenue generated from these two sources are shown separately. It is quite evident from the figure that revenue generated from customs duties has indeed increased significantly in absolute terms over the past years. In 2003-04, it has crossed 1200 million US$ mark which was 915 million US$ and 650 million US$ in 1993-94 and 1989-90 respectively. Revenue generated from other duties is available from 1991-92, and the indication is that its contribution has been increasing in the total trade tax. In 2003-04, more than 1000 million US$ worth of revenue was collected from other trade duties which was only 339 million US$ in 1991-92.

However, relative contribution of trade generated revenue was quite stable in second and third phase, trade tax constituting about three to four per cent of GDP; nonetheless, previously, before 1991-92, this was below 2 per cent (revenue figures from other duties is not available for this period).

Interestingly, during the same period contribution of total tax to GDP has shown an increasing trend and reached 8.98 per cent in 2003-04. Although the contribution of tax revenue is very low for Bangladesh and given the fact that contribution of trade tax was rather stagnant during trade liberalisation period, rising involvement of tax revenue to GDP indicates that liberalisation has not hurt revenue generation *per se*.

Figure 3.11

Trend in Customs Duty, Para Tariff and Trade Tax and Tariff Revenue as Per cent of GDP

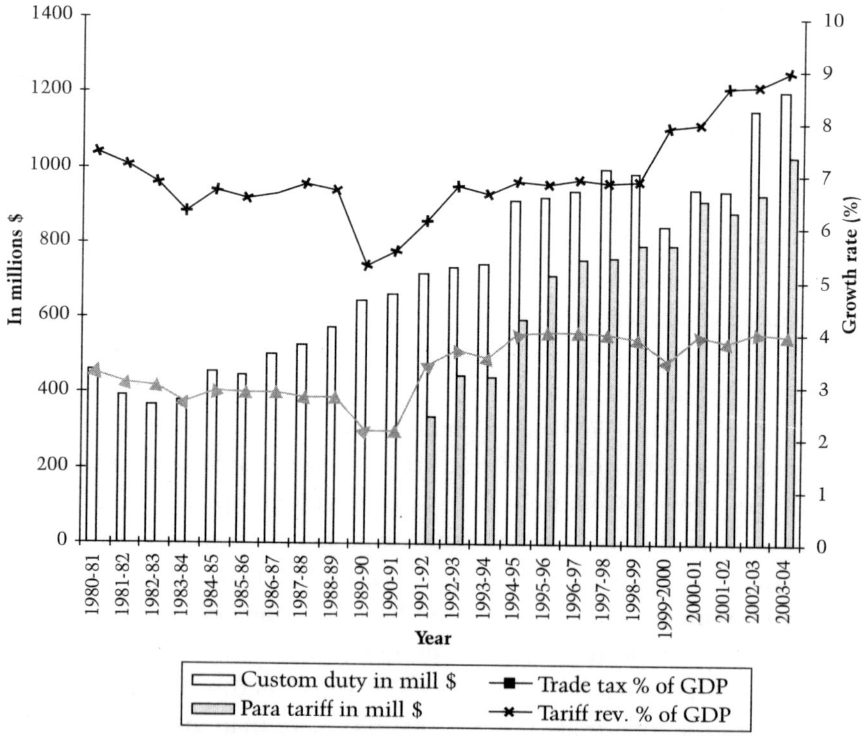

Source: Authors' estimation based on data collected from the various issues of BER.

f) Terms of Trade (ToT): ToT is measured by the ratio of export price index to import price index. Although trade liberalisation does not have a direct bearing on ToT, in an indirect way it does have as it shows the relative price of export and imports which in turn are influenced by trade policies. Figure 3.12 shows trends in terms of trade by making use of export and import price indices.

Figure 3.12

Trend in Terms of Trade (TOT)

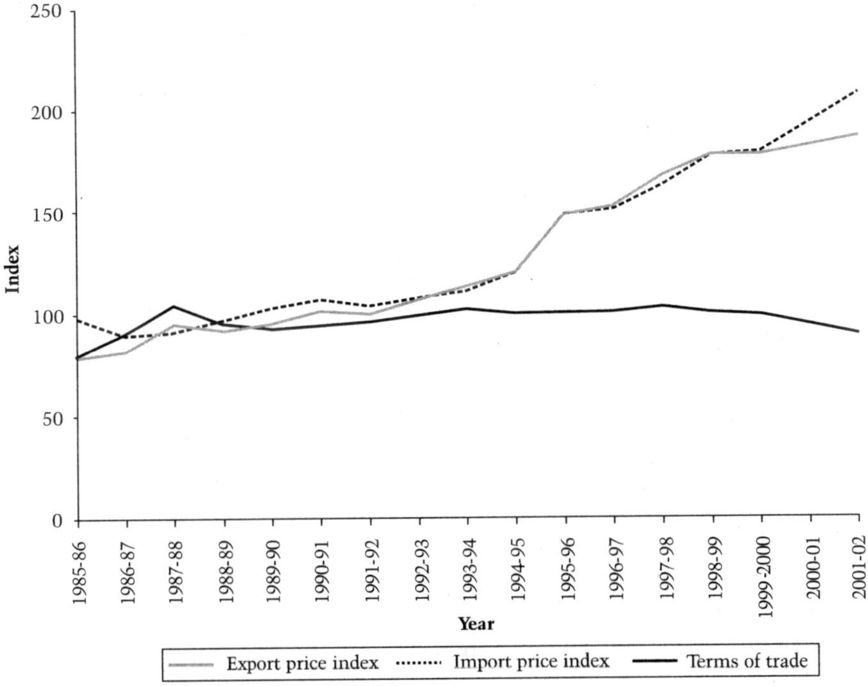

Source: Authors' estimation based on data collected from the various issues of BER.

Figure 3.12 shows that starting from late 1980s, both export and import price indices have increased; however, growth in import price index was higher for most of years resulting in a deteriorating terms of trade situation for Bangladesh over these years. During the second phase, an increasing trend is seen indicating that more imports can be bought by the same amount of exports. However, from 2000 onwards a declining trend is observed in the country's terms of trade.

3.2.3. Policy Related Indicators

a) Real Effective Exchange Rate (REER): Real effective exchange rate implies how much domestic currency is depreciated against a foreign currency. If domestic currency is depreciated (to a significant extent) against foreign currency, it is expected to reduce anti-export bias in the trading system. Table 3.5 shows trends in two REER indices for Bangladesh.[5]

Table 3.4

Trend in REER Index

(a) CPD, Base Year 1991=100

Year	*1991*	*1992*	*1993*	*1994*	*1995*	*1996*	*1997*	*1998*	*1999*	*2000*
REER	100	100.2	104.6	102.2	102.9	103.8	105.5	105.4	112.8	116.2

Source: CPD-IRBD (2002).

(b) Bangladesh Bank

Fiscal Year	*1996*	*1997*	*1998*	*1999*	*2000*	*2001*	*2002*	*2003*	*2004*	*2005*	*2006*	*2007 (end of March)*
REER Index	95.5	102.2	109.6	111.6	108.1	102.0	101.5	97.0	93.4	91.7	86.9	84.6

Source: Bangladesh Bank.

Table 3.4(a) shows that from early 1990s onwards the REER has increased till 2000 indicating that domestic currency has depreciated more with respect to other currencies. However, Table 3.4(b) reveals that from 2001 the trend in REER has reversed; the domestic currency has appreciated compared to other currencies.

b) Anti-Export Bias: Anti-export bias is the ratio of effective exchange rate for import and export. If there is anti-export bias in the economy, it could help import substituting sectors, but harm export-oriented sectors. During the pre-reform phase, anti-export bias was prominent in trade policy because of high import duties and high dispersion. However, with liberalisation of trade, there was a change. In 1992, 1993 and 1994, the anti-export bias increased because of the increase in import weighted

5. These indices are found from two different sources and covering different time periods.

CD+SD+RD (custom duty, supplementary duty and regulatory duty) and protective effects of supplementary duties.[6] However, since the late 1990s, the bias has started to decrease, as is shown by Figure 3.13.

Figure 3.13

Trend in Effective Exchange Rate of Import and Effective Exchange Rate of Export and Anti-Export Bias during 1990s

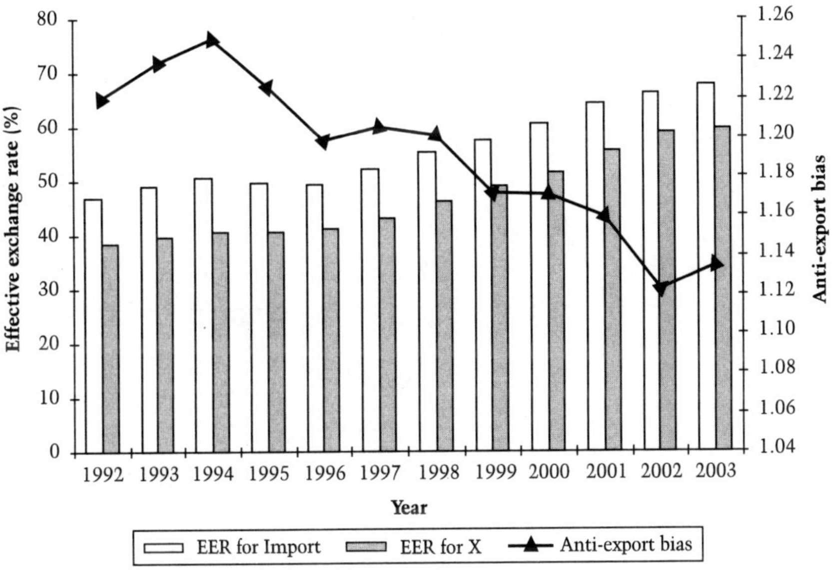

Source: Reproduced from Taslim (2004).

This declining anti-export bias has favoured the export sector (ratio above 1 indicates bias against export).[7] As a World Bank study indicates, the bias is more significant for final consumer goods (World Bank, 1999). Since 1999, the anti-export bias has fallen further, with an insignificant rise in 2003.

---

6. Rahman, M. and Raihan (2007). *Bilateral Trade Deficit of Bangladesh: Is it a Matter of Concern?*
7. Bhattacharya, D and Titumir. "Bangladesh Experience with Structural Adjustment: Learning from a participatory Exercise." Here they mentioned that the World Bank estimate is likely to give an overestimate of anti-export bias due to not taking into account import taxes on imports.

c) Black Market Premium: The black market premium shows the extent of over-valuation of the official exchange rate. With trade liberalisation the black market premium is expected to decrease over time. This decrease is expected to stimulate activities in the tradable sectors.

Figure 3.14

Trend in Black Market Premium (Per cent)

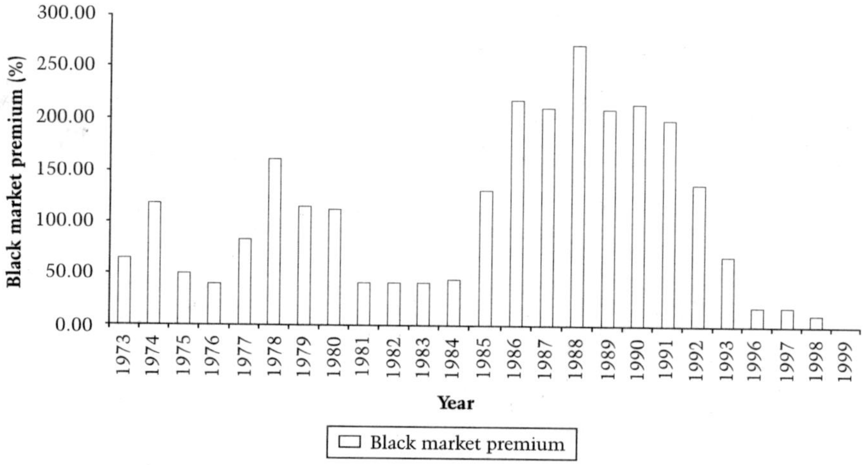

Source: World Bank (2006).

In early 1980s, the level and fluctuations in black market premium was quite high; this was the picture in the decade of 1980s and continued in 1991 and 1992. From 1996, it started to decrease; as a matter of fact, the premium has been zero since 1999 when the secondary market was abolished and the dual exchange rate was modified (Figure 3.14).

3.3 Sequencing of Trade Liberalisation

According to Falvey and Kim (1992), timing and sequence is important in designing a trade liberalisation process. Levine and Renelt (1992) pointed out that policies correlated with growth (trade openness, macro stability, small government consumption, rule of law) are all highly correlated among themselves. Therefore, when all of these policies are included in the regression analyses, it may be difficult to identify individual effects of the various policies. Moreover, it is easy to mis-attribute the effects of omitted polices and institutional variables in this process. According to Winters *et al.* (2004), to have a long-lived or permanent effect on growth, trade liberalisation almost certainly requires combination with other appropriate policies such as encouraging investment, allowing effective conflict resolution and promoting human-capital accumulation. Government can take trade liberalisation policy at one go or can initiate it after taking into account several other policies in the areas of macroeconomic stability, capital market perfections, adjustment costs, quota restrictions, tariffs, etc.

In case of Bangladesh, a number of other policies have also undergone important changes during the different phases of trade liberalisation. In the next few sections, some of these policies will be discussed to provide a better understanding about the sequencing of trade liberalisation policies.

3.3.1 Import Policies

During the protectionist regime (1972-1980), Bangladesh followed an extremely restrictive import policy. Major administrative instruments employed to implement the import policy during this time were the foreign exchange allocation system and the Import Policy Orders (IPOs). Under the IPOs, items were specified accordingly to whether their importations were allowed, prohibited or required special authorisation. The government relied upon import-licensing rather than on tariffs to ensure allocation of foreign exchange to priority areas, and to protect the local industries from import competition.

During the first phase of trade liberalisation (1981-1990), the pace of import liberalisation was moderate. In 1984, import-licensing system was abolished and imports were permitted against letters of credit (L/C). Since 1986, there had been significant changes in the import procedures and in the IPOs with respect to their contents and structures; no restrictions were imposed on import of items that did not appear in the IPOs (Ahmed, 2001 cited in SR, 2007). From 1986, a Negative List (for banned items) and the Restrictive List (for items importable on fulfilment of certain prescribed conditions) was introduced. In 1990, the negative and restricted lists of importable were merged into one 'Consolidated List'.

During the second period, i.e., from 1991, the pace of import liberalisation was rapid. In addition, the tariff was reduced alongwith reductions in the number of rates and compressing tariff bands. As a result, average nominal tariff rates of protection for all tradables fell from 88.6 per cent in 1991 to 22.2 per cent in 1999. Import-weighted tariff and dispersion in tariff rate were also reduced. As a result of the continued liberalisation of the import regime, the number of tariff slabs was reduced from 8 in the pre-reform time to 3 in the most recent period (2007-08 budgets); maximum tariff rate has also come down quite drastically as was mentioned earlier.

3.3.2 Export Policies

Since the early 1980s a number of reforms were implemented in the areas of duties on inputs for exported items, exchange rate, monetary and fiscal policy incentives which were aimed at increasing effective assistance to exports. RMG sector has been the major beneficiary of these reforms. Special Bonded Warehouse Scheme was introduced early in 1978 allowing exemption from import taxes on inputs for export; a Duty Drawback Scheme was also introduced for the RMG sector. Various incentives were provided to exporters including the unrestricted and duty-free access to imported inputs, financial incentives in the form of easy access to credit and credit subsidies and a number of fiscal incentives such as rebates on income taxes and concessionary duties on imported capital machinery.

Until 1986, under the Export Performance Licensing Scheme, exporters of non-traditional products used to be allocated import licences for specific products over and above their allotment. In 1986, the Export Performance Licensing (XPL) Scheme was replaced by the Export Performance Benefit (XPB) Scheme providing an added benefit to exporters by allowing them to cash it in the secondary exchange market. Back-to-Back L/C System was also introduced in 1987 that enabled exporters to import raw materials on a deferred payment basis. Export Credit Guarantee Scheme in 1978 and Cash Compensation Scheme in 1986 were introduced. The Export Promotion Fund was established to provide exporters venture capital on easy terms and lower interest rates. Various fiscal incentives such as rebates on income taxes and concessionary duties on imported raw materials and capital machineries were also put in place. Export Promotion Bureau (EPB) was established with the aim of promoting exports and designing plans and policies conducive to the private sector.[8]

3.3.3 Industrial Policy

In the period immediately after liberation in 1971, the government of Bangladesh nationalised all the heavy industries, banks and insurance companies; by 1972 the nationalised units accounted for 92 per cent of the total fixed assets of the manufacturing sector in Bangladesh. Such a nationalisation process restricted private sector participation only to the medium, small and cottage industries. After 1975, the government moved away from the nationalisation programme and revised the industrial policy with a view to facilitating a greater role for the private sector. The new industrial policy of 1982 (NIP-82), and the revised industrial policy of 1986 (RIP-86) were important milestones in this respect. The ceiling on investment was raised and subsequently abolished, the positive list was replaced by the negative list and a number of publicly managed industrial enterprises were denationalised and transferred to private ownership. EPZs were set up to assist export-oriented industries. Plans are there to set up industrial parks and special economic zones. Small and medium enterprises (SMEs) was identified as a priority sector; various measures were taken to support, with varying degree of success.

In the present Industrial Policy, four reserved sectors have been identified and investments in the public sectors are restricted to these reserved sectors only. To attain the industrial growth, special importance was given in the Industrial Policy on agro-based and agro-processing industries and on steps to overcome possible adverse conditions in the

---

8. In the Proposed Export Policy 2006-2009, the number of thrust sectors from export policy 2003-2006 was increased to 6 from 5 and Special Development Sectors decreased to 9 from 10. The 6 newly approved thrust sectors are: i. Agro-products and agro-processing products/goods; ii. Light engineering goods/products (including auto-parts and bicycles); iii. Shoe and leather goods/products; iv. Pharmaceuticals products/goods; v. Software and ICT products/goods; vi. Home textiles.

   The newly approved 9 special development sectors are: i. Finished leather production; ii. Frozen fish production and processing; iii. Handicrafts; iv. Electrical goods; v. Fresh flower and fruits; vi. Jute goods; vii. Handloom from hill tracts.

export-oriented garment sector. The current Industrial Policy identified 33 industries as thrust sectors for which various incentives were given. The trade reforms pursued during this time were seen as a necessary complimentary process which would support the industrial policy. Lowering of import duties for industrial inputs and machineries, tax holidays and taxation at a reduced rate or accelerated depreciation were some of such incentives. It was decided that there will be no differences in terms of incentives between the same types of industries established in the public and private sectors in respect of customs duty and other taxes. The duty structures of imported raw materials, intermediate goods and manufactured goods were set at gradually escalating rates to support the industrial policy. The 2005 Industrial Policy came up with a number of new fiscal and financial measures to support the thrust sectors. Alongwith the above mentioned steps, several other revenue and financial measures were accorded importance in the Industrial Policy 2005.[9]

3.4 Comparison with Rest of the World

Table 3.5 compares average tariffs in pre-reform and post-reform periods for Bangladesh and some selected regions of the world.

For Bangladesh, during phase 2, the post-reform average tariff rate was low compared to South Asian and African countries, but was high compared to Latin American or East Asian countries. In phase 3, the speed of liberalisation slowed down in Bangladesh (compared to other regions). Table 3.5 shows that the tariff ratio (phase two over phase 1) was lowest in case of Bangladesh compared to other regions of the world indicating the rapid pace of falling MFN tariffs during that particular period (indicating faster pace of trade liberalisation in Bangladesh). It is also to be noted that in comparison to other regions, the pace of reduction in MFN tariff of Bangladesh was lower (phase 3 over phase 2 as indicated by tariff ratio of 0.66) indicating slowdown in reduction in tariffs in Bangladesh in recent years.

9. For details, see Industrial Policy 2005.

Table 3.5

Comparisons of Tariff Regimes with the Rest of the World

Region Averages	*Average MFN Tariff*			*Tariff Ratio**	
	Phase 1	*Phase 2*	*Phase 3*	*Phase 2 over Phase 1*	*Phase 3 over Phase 2*
Bangladesh	94 (1989)	23 (1998)	15.2 (2006)	0.24	0.66
South Asia	80 (1985)	27 (1998)	14.2 (2006)	0.34	0.53
East Asia	29 (1984)	25 (1992)	6.98 (2006)	0.86	0.28
Africa	41 (1983)	38 (1993)	na	0.93	Na
Latin America	41 (1984)	15 (1993)	8.72 (2006)	0.37	0.58

Note: *Tariff ratio is the post-reform average tariffs over pre-reform average tariffs.
Source: World Tariff Profile (2006); World Bank (1994) and Rajapatrirana (1997).

Table 3.6

Trend in Simple Average Tariff Rates in South Asia (1998-99 to 2005-06)

Countries	*1998-99*	*2005-06*
Bangladesh	20	15.2
India	39.6	19.2
Pakistan	21.3	14.3
Sri Lanka	17.6	11.2
Nepal	14	13.9

Source: World Bank, 2004; World Tariff Profiles, 2006.

From a regional perspective, except for Pakistan, Sri Lanka and Nepal, the average MFN tariff rate was lower in Bangladesh. However, this does not give a clear picture about the relative restrictiveness of individual countries since it does not include para-tariffs (Table 3.6).

Figure 3.15 shows the degree of openness for some selected neighbouring countries. It can be seen from the figure that barring Pakistan, Bangladesh trade orientation was lower compared to all other countries.

Figure 3.15

Dynamics of Trade Openness for Selected Countries

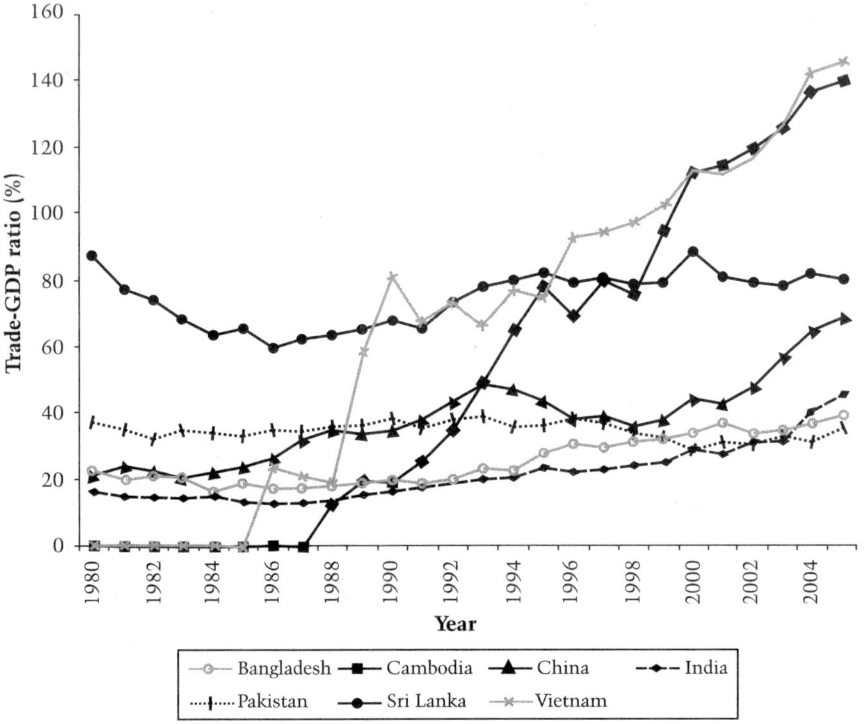

Source: WDI CD-ROM, 2007.

3.5 Summary

It may be concluded from the aforesaid review of trade reforms, export and import policies, and industrial policies and the related indicators, that since the 1980s, Bangladesh economy has undergone significant changes in terms of degree of protection, removal of anti-export bias, effective rates of protection to domestic industries and incentives provided to export-oriented industries. Both trade and industrial policies have changed quite significantly over the past years. The analysis in this chapter indicates that the policy of trade liberalisation was supported by a host of reforms in other areas which have gone a long way in creating a conducive environment for promotion of export-oriented and private sector-led activities in the economy. What implications this combination of policies had in terms of employment, real wages and productivity is of crucial importance from the perspective of capturing the real worth of these major reforms. This is understandably critically important for Bangladesh from policy perspectives. The impact study carried out in the next four chapters will attempt to answer some of the relevant questions in this context.

Annex A-3.1

Notes on the Definition of Indicators

1. Trade dependency ratio or Trade-GDP ratio

Trade dependency ratio is defined as the ratio of exports and imports to GDP, i.e., imports plus exports as a share in GDP.

Mathematically, trade dependency ratio = (Trade/GDP)*100

2. Growth rate of exports

Growth rate of exports is defined as the growth rate of exports over the specified period.

Mathematically, growth rate of exports = $\frac{E_c - E_b}{E_b} * 100$,

Where E_c = Exports in the current year

E_b = Exports in the base year

3. Average tariff rate

The simple average tariff rate is the MFN applied duty rate averaged over all the tariff lines on the goods imported into the country. It is the direct measure or the policy measure of trade liberalisation.

The trade-weighted average tariff rate is the import weighted average MFN applied tariff rate at the tariff line level; i.e., the nominal MFN applied tariff rate at each tariff line is weighted by the import share of the respective tariff line in the total import and then taking their average gives the trade or import weighted tariff rate.

4. Quantitative restriction/Non-tariff barriers (NTB) on import

There are various forms of non-tariff measures, such as quotas, import-licences, voluntary export restraints and export taxes. None of the average tariff measures are able to capture any of these restrictions or impediments to trade affecting both imports and exports. These non-tariff barriers (NTB) can have a much greater effect upon trade flows and these are most widely used as quantitative restrictions (QRs) on imports.

5. Black market premium

The black market premium is the difference between the official exchange rate and the rate in the secondary market. This premium shows the extent to which the official exchange rate is overvalued. The unification of the exchange rate or the abolition of black market premium is one of the major objectives of trade liberalisation.

Black market premium can be calculated as follows:

i.e., Black market premium = Official exchange rates – Secondary exchange rate

If the difference is high (the premium is high) then it can affect both exports and imports. When it is high, it indicates that the local currency is over-valued. As a result, exports become uncompetitive and the demand for imports decreases, as it appears to be a better option not to spend the scarce foreign currency.

6. Anti-export bias

The index of anti-export bias is defined as the ratio of the effective exchange rate of imports (EER_m) to the effective exchange rate of exports (EER_x).

Mathematically it is defined by, anti-export bias $= \dfrac{EER_m}{EER_x}$

If the ratio is greater than one then the trade regime in a particular country is biased against exports.

7. Import weighted average tariff rate

Import weighted average tariff rate measure weights each tariff by the share of total imports in that import category. It has been widely used in the trade literature to evaluate the extent of trade protection in a country.

8. Implicit nominal tariff rate

The implicit nominal tariff rate is the ratio of collected revenue from import duties to the value of imports at the c.i.f. values. Thus, the implicit or *ex post* nominal tariff rate can be calculated as:

$$tp = \frac{\Sigma TP}{\Sigma PQ} \quad \text{Where } \Sigma TP = \text{Total customs collection}$$

$$\Sigma PQ = \text{Total value of imports}$$

Implicit nominal tariff is a more reliable measure of trade restriction than the simple average tariff because the implicit nominal tariff rate is implied by the actual collections of custom revenues. It takes into account the duty-free and preferential-rate imports.

9. Effective rate of protection

The effective rate of protection (ERP) takes into account the total effect of tariffs on input and output in a production process. Effective rate of protection can be defined as follows:

$ERP_j = \frac{t_j - a_{ij}t_i}{1 - a_{ij}}$ Where, ERP_j = Effective protection rate for activity j

t_j = Nominal tariff rate on activity j

a_{ij} = Share of activity I in cost of activity j at free trade price

t_i = Nominal tariff on activity I

10. Import-penetration ratio

The import-penetration ratio is the share of total imports in GDP. It can be calculated as follows:

Import-penetration ratio = (Total import/GDP)*100

A low penetration rate may also reflect the presence of highly competitive domestic firms capable of resisting foreign competition, especially if the export ratio is high at the same time. Conversely, a high import-penetration rate may reflect weak competitiveness of domestic firms, especially if the export ratio is low. If both indicators are high that reflects internationalisation of industries, especially relating to the sourcing of intermediate goods, intra-industry trade and intra-firm trade.

11. Export-orientation ratio

The export-orientation ratio is the ratio of a country's total exports to its GDP. It can be calculated as follows:

Export-orientation ratio = (Total export/GDP)*100

12. Import-penetration of consumer goods

The import-penetration of consumer goods is defined as the share of the import of consumer goods in GDP. It calculated as follows:

Import-penetration of consumer goods = (Total imports of consumer goods/GDP)*100

13. Real effective exchange rate

The changes in the real effective exchange rate show the extent in which local currency is depreciated against the foreign currency. It can be calculated as follows:

$$\text{Real effective exchange rate} = \frac{E_c - E_b}{E_c} * 100$$

Where E_c = Exchange rate in the current year

E_b = Exchange rate in the following year

Annex Table A-3.1

Tariff Reform Indicators

Year	*Simple (Unweighted) Average Tariff (%)*	*Import Weighted Average Tariff Rate (%)*	*Implicit Nominal Tariff Rate (%)*
1973	98	n/a	n/a
1974	98	n/a	n/a
1975	98	n/a	n/a
1976	98	n/a	n/a
1977	98	n/a	19.2
1978	98	n/a	18.1
1979	98	n/a	18.8
1980	98	n/a	15.9
Average for Pre-Reform Period	98	n/a	18
1981		n/a	16.2
1982		n/a	13.8
1983	100	n/a	14.1
1984		n/a	14.9
1985		n/a	16.2
1986	82	n/a	17.5
1987		n/a	17.7
1988	104	n/a	16.9
1989	115	n/a	15.8
1990		n/a	16.1
Average for Phase 1 of Trade Liberalisation	100.25	n/a	15.92
1991	88.6	42.1	17.7
1992	57.5	24.1	18.1
1993	47.4	23.6	16.4
1994	36	24.1	18.2
1995	25.9	20.9	16.5
Average for Phase 2 of Trade Liberalisation	51.08	26.96	17.38
1996	22.30	17.00	14.60
1997	21.50	18.00	14.62
1998	20.70	16.00	14.93
1999	20.30	14.10	14.31
2000	19.50	13.80	12.35
2001	18.60	15.10	12.62
2002	17.13	9.73	13.71
2003	16.51	12.45	14.22
2004	15.62	11.48	13.65
2005	13.54	9.59	n/a
2006	13.41	8.09	n/a
2007	12.24	6.98	n/a
Average for Phase 3 of Trade Liberalisation	17.61	12.69	13.89

Source: BER, NBR, BTC.

Annex Table A-3.2

Trade Related Indicators

Year	Growth Rate of Exports	Degree of Openness	Export Penetration Ratio	Export Penetration Ratio of Manufacturing Goods	Import Penetration Ratio	Import Penetration of Consumer Goods	Trade Tax % of GDP	Terms of Trade
1973	n/a	n/a	n/a	n/a	n/a	n/a	n/a	n/a
1974	6.70	n/a	n/a	n/a	n/a	n/a	n/a	n/a
1975	2.94	n/a	n/a	n/a	n/a	n/a	n/a	n/a
1976	-0.58	n/a	n/a	n/a	n/a	n/a	n/a	n/a
1977	9.60	n/a	n/a	n/a	n/a	1.70	n/a	n/a
1978	18.40	n/a	n/a	n/a	n/a	1.80	n/a	n/a
1979	25.33	n/a	n/a	n/a	n/a	2.01	n/a	n/a
1980	21.11	24.6073	5.92	4.04	18.69	3.00	3.23	n/a
Average of Pre-Reform	11.93	24.61	5.92	4.04	18.69	2.13	3.23	
1981	-5.28	22.7849	4.99	3.590874	17.80	2.03	3.049147	n/a
1982	-11.83	24.7786	4.85	3.358814	19.93	2.60	2.988545	n/a
1983	9.70	24.2425	5.55	3.650702	18.69	2.92	2.684507	n/a
1984	18.12	22.1391	5.69	3.724767	16.44	2.90	2.90738	n/a
1985	15.22	22.8540	5.96	3.942516	16.89	3.95	2.872338	n/a
1986	-12.33	20.4140	5.25	3.333719	15.16	3.20	2.85957	80.1
1987	31.42	20.9991	6.12	4.408175	14.88	3.10	2.770556	91
1988	14.36	22.0627	6.44	4.943866	15.62	4.20	2.793875	104.7
1989	4.90	22.7385	6.29	4.828796	16.45	3.80	2.130172	95.3
1990	17.97	17.3336	5.00	3.940727	12.33	3.02	2.137833	92.8

contd...

...contd...

Year	*Growth Rate of Exports*	*Degree of Openness*	*Export Penetration Ratio*	*Export Penetration Ratio of Manufacturing Goods*	*Import Penetration Ratio*	*Import Penetration of Consumer Goods*	*Trade Tax % of GDP*	*Terms of Trade*
Average of Phase 1	8.23	22.03	5.61	3.97	16.42	3.17	2.72	92.78
1991	12.72	16.8744	5.54	4.555319	11.33	3.70	2.297117	94.9
1992	16.09	17.6162	6.36	5.508265	11.25	4.20	2.293805	96.2
1993	19.51	20.1509	7.44	6.461035	12.71	4.15	2.203335	99.6
1994	6.34	19.8690	7.49	6.462419	12.38	3.69	2.410824	102.3
1995	37.04	24.5313	9.15	7.962614	15.38	5.01	2.268211	100.1
Average of Phase 2	18.34	19.81	7.20	6.19	12.61	4.15	2.29	98.62
1996	11.80	26.4311	9.54	8.372095	16.89	4.92	2.220823	99.94
1997	15.22	27.3440	10.44	9.220478	16.90	4.08	2.267543	101.12
1998	16.80	28.7967	11.72	10.60552	17.08	4.70	2.156943	103.09
1999	2.94	29.1413	11.63	10.72769	17.52	5.26	1.794218	100.02
2000	8.20	29.6719	12.20	11.2158	17.47	4.46	2.011741	99.11
2001	12.52	33.7558	13.77	12.73526	19.99	4.53	1.974766	94.39
2002	-7.44	30.5774	12.59	11.76753	17.99	3.69	2.22218	89.85
2003	9.33	31.1963	12.61	11.72331	18.59	4.40	2.12856	n/a
2004	16.10	32.6286	13.44	12.4696	19.19	4.58	2.134529	n/a
2005	11.01	35.9147	13.94	13.23546	21.97	5.27	1.882296	n/a
2006	24.65	40.7140	16.95	15.72383	23.76	5	n/a	n/a
Average of Phase 3	11.01	31.47	12.62	11.62	18.85	4.63	2.08	98.22

Source: BER, BBS, Bangladesh Bank, EPB.

Annex Table A-3.3

Policy Related Variables

Year	Real Effect Exchange Rate (REER)	Anti-Export Bias = EER_m/EER_x	Black Market Premium
Unit	Base Year 1991=100		
1973	n/a	n/a	64.63
1974	n/a	n/a	117.82
1975	n/a	n/a	50.71
1976	n/a	n/a	39.80
1977	n/a	n/a	83.33
1978	n/a	n/a	161.22
1979	n/a	n/a	114.19
1980	n/a	n/a	112.31
Average of Pre-Reform	n/a	n/a	93.00
1981	n/a	n/a	41.06
1982	n/a	n/a	41.25
1983	n/a	n/a	42.00
1984	n/a	n/a	45.19
1985	n/a	n/a	130.31
1986	n/a	n/a	218.18
1987	n/a	n/a	210.90
1988	n/a	n/a	271.86
1989	n/a	n/a	209.89
1990	n/a	n/a	214.83
Average of Phase 1			142.55
1991	100.00	n/a	199.21
1992	100.20	1.22	135.98
1993	104.60	1.24	67.23
1994	102.20	1.25	n/a
1995	102.90	1.23	n/a
Average of Phase 2	101.98	1.23	134.14
1996	103.80	1.20	18.66
1997	105.50	1.21	19.15
1998	105.40	1.20	11.11
1999	112.80	1.17	0
2000	116.20	1.17	n/a
2001	n/a	1.16	n/a
2002	n/a	1.12	n/a
2003	n/a	1.13	n/a
Average of Phase 3	108.74	1.17	12.23

Source: BTC, FRBD, WB 2002.

Annex Table A-3.4

Average Import Tariff Rates for Various Categories of Imported Good

Year	*Category*	*Primary Product*	*Intermediate Product*	*Capital Product*	*Final Consumer Product*
1991-92	Unweighted	55.22	49.51	45	80.29
	Weighted	23.37	24.14	18.71	47.27
1992-93	Unweighted	47.74	41.63	38.68	63.56
	Weighted	23.31	23.65	18.46	36.47
1993-94	Unweighted	34.89	31.8	26.16	49.89
	Weighted	27.15	22.87	16.15	36.68
1994-95	Unweighted	31.16	23.23	13.87	38.62
	Weighted	17.26	26.26	12.51	26.45
1995-96	Unweighted	24.55	20.15	12.3	33.69
	Weighted	13.16	22.65	9.54	24.09
1996-97	Unweighted	22.63	19.6	12.5	32.3
	Weighted	16.62	22.2	10.4	23.1
1997-98	Unweighted	21.9	19.2	12.1	30.6
	Weighted	13.6	21.3	8.2	20.1
1998-99	Unweighted	21.4	19	12.3	29.2
	Weighted	9.5	21.3	8.1	17.6
1999-2000	Unweighted	18.3	15.2	8.2	26.9
	Weighted	10.7	18.6	7.7	17.7
2000-01	Unweighted	20.1	15.8	7	25.7
	Weighted	9.8	16.8	6.2	15.8
2001-02	Unweighted	20.1	15.61	6.97	26
	Weighted	9.43	16.18	3.26	13.96
2002-03	Unweighted	20.98	14.89	8.03	22.94
	Weighted	11.92	15.86	7.97	11.72
2003-04	Unweighted	19.9	14.44	7.85	21.27
	Weighted	11.28	15.12	6.42	10.68
2004-05	Unweighted	17.61	12.46	7.28	18.22
	Weighted	8.99	12.72	5.22	15.08
2005-06	Unweighted	17.83	12.24	7.45	18.13
	Weighted	6.93	9.33	5.16	13.44

Source: Various issues of BER.

4 Impact of Trade Liberalisation in Bangladesh at the Aggregate Level

An Analysis of LFS Data

During the various phases of trade liberalisation, significant changes have taken place in the overall employment situation and with respect to various characteristics of the labour force. It is, however, rather difficult to disentangle the singular impact of trade liberalisation on the structural changes in the labour force from other factors acting in tandem in the economy. The focus of the analysis in this study will be on the employment scenario of Bangladesh which has been presented in this chapter. This chapter has analysed detailed reports of the six Labour Force Survey (LFS) covering the period between 1980 and 2002 along with the provisional summary findings of the latest LFS 2005-06. It is to be noted in this context that successive LFS reports are major sources of employment data in Bangladesh. As part of the analysis, different indicators of employment were traced and compared over the various phases of trade liberalisation in Bangladesh. The years have been chosen in such a manner that each trade liberalisation phase covers two periods.[1] The next chapter will deal with dynamics of manufacturing employment and a range of key industrial statistics (based on CMI data) in relation to the various phases of trade liberalisation and the pre-reform period. Regrettably, both the aforesaid datasets (LFS and CMI) have serious limitations. It is critically important for the present study that an attempt is made to correct and adjust the data provided by these sources in order that a comparison could be made for the period under review. This exercise will also be of much help to future

1. Final LFS reports of 2002-03, 1999-2000, 1995-96, 1990-91, 1989 and 1983-84 have been utilised in this analysis. The preliminary findings of the latest LFS 2005-06 has also been reported where possible.

researchers who would like to undertake research in related areas. Before proceeding with the analysis itself, the formidable work undertaken in this respect has been discussed first.

4.1 Problems with LFS and CMI Data

The limitations of the data reported by BBS, particularly on employment statistics, are rather well known to Bangladeshi audience. These limitations have severely constrained the work of researchers dealing with Bangladesh's employment situation. One major difficulty has been that data for various reports is difficult to compare. For LFS, data are not of panel type in nature; rather the survey is of cross-section type by nature limiting the scope for more in-depth analysis. The changes in the definition of the civilian labour force (extended *versus* usual definition) and in age criterion for economically active population (10 years+ *versus* 15 years+) makes it difficult to compare the various labour force indicators provided in successive LFS reports.[2] This has given rise to confusion and debate among researchers.[3] For the purpose of our all analyses the usual definition of economically active population that included 15 years+ population group has been consistently used in the present study. The LFS reports of 1983, 1989 and 1990-91 have used 10+ population group as the definition for economically active population while LFS report of 1990-91 have reported extended definition of economically active population. In the report of LFS 2000, analysis based on both extended and usual definition and 10+ and 15+ age group were reported for 1995-96 and 2000 labour force data. It is to be noted here that recent years (LFS 2002-03, and 2005-06) have used the usual definition and 15+ population group. We have taken this as the

---

2. For the definition of indicators of LFS please see Annex A-4.1.
3. According to usual definition, any person aged 10 years and above who was either employed or unemployed during the reference period and any person of the same age putting in a minimum of one hour's work in family farms/enterprises for pay or profit during the reference period is considered economically active. On the other hand, according to extended definition, any person of the same age specification who was either employed or unemployed during the reference period and who was engaged in household economic activities such as care of poultry and livestock, threshing, cleaning, boiling, drying, processing and preservation of food etc., with or without pay or profit during the reference period is treated as economically active (LFS, various issues).

reference point for bringing consistency both in terms of definition (usual) and also the age cohort for economically active population (15+). Based on the proportions reported in LFS 2000 and using the data on age structure of the population, an attempt has been made in this study to adjust the data of various LFS so that the labour force statistics can be compared in a consistent manner.

Table 4.1

Adjustment Factors for LFS Data

Adjustment Factors	*1983*	*1989*	*1990-91*
For correcting age criteria (From 10+ to 15+)	0.9690	0.9701	0.9695
For correcting definition (From extended to usual definition)	-	-	0.69

Source: Authors' estimation based on LFS and BBS data.

In addition, the manufacturing industry data presented in the CMI, particularly at the disaggregate level, suffers from inconsistency and lack of reliability. A close examination of disaggregated level data indicates that data for a significant number of variables (in particular years) is rather of doubtful nature. The raw data, which has been cleaned using various outlier identification techniques, was used for a number of variables and indicators.

Even if one concedes that the quality of data for industries is poor, the state of data for agriculture is still poorer. Unavailability of disaggregate level employment data for agriculture did not allow an exercise of the type that was undertaken for the industrial sectors based on CMI data.

4.2 Dynamics and Structural Changes in the Labour Market during Various Phases of Trade Liberalisation

This analysis of the labour force of Bangladesh begins with an overview of labour market scenario depicting trends in population, civilian labour force (CLF), employment and labour force participation rate (LFPR). From Table 4.2, it is evident that LFPR has experienced a positive growth from year to year. From a very low LFPR of 44.5 per cent in 1983-84, it jumped to reach 47 per cent at the end of phase one (1990-91) which increased further by five percentage points at the end of phase two (i.e., in

1995-96). One striking feature is that 15+ population group in 1990-91 was 71.0 million which has dropped to 69.5 million in 1995-96, despite the positive growth in overall population. Since then, the population size of 15+ age group has shown a persistent increasing trend. Preliminary findings of LFS 2005-06 has reported LFPR as 58.5 per cent which is only 1.2 percentage point higher than what was reported in the previous LFS of 2002-03. The average per annum growth rate for LFPR was highest during second phase (0.98 per cent) followed by third phase (0.93 per cent) and first phase (0.92 per cent) respectively (Figure 4.1).

Table 4.2

Trend of Civilian Labour Force, Labour Force Participation Rate and Employment

	Phase 1		*Phase 2*		*Phase 3*		
	1983-84	*1989*	*1990-91*	*1995-96*	*2000*	*2002-03*	*2005-06*
Population (15+), in million	62.9 (64.9)	68.7 (70.8)	71.0 (73.5)	69.5	74.2	80.8	84.6
Civilian labour force (15+), in million	28.0 (28.5)	32.3 (33.3)	35.0 (35.9)	36.0	40.7	46.3	49.5
Total employment, in million	27.0 (28.0)	32.0 (32.7)	34.0 (34.9)	34.8	39.0	44.3	47.4
Labour force participation rate (LFPR), in %	44.5 (43.9)	47.0 (47.03)	48.8	51.8	54.9	57.3	58.5

Note: (1) Students, disabled persons, armed services personnel, inmates of jails, pensioners and other income recipients etc., are excluded from the civilian labour force.

(2) Labour force participation rate is estimated as - total civilian labour force divided by total population of 15+ ages.

(3) All figures in parentheses indicate the figure reported in respective LFS reports which are based on 10+ age criteria.

Source: Authors' estimation based on data collected from the various issues of LFS and BBS.

Figure 4.1 reveals some quite interesting scenario as regards the (phase-wise) average growth in the overall employment situation of Bangladesh. Average population growth has remained stagnant at around 2.2 per cent during first and second phases, which fell significantly to 1.42 per cent in the third phase. The mean growth rates, both for total civilian labour force and employment, were over 2 per cent in the first and third phases whereas these variables were unusually low in the second phase, 0.47 per cent and 0.39 per cent respectively. However, these growth rates for the second and third phases, as reported in LFS, did not conform to the population growth during the corresponding period. Observing the sectoral and overall employment growth, it can be stated that during the second phase of trade liberalisation, employment, in general, has suffered. In particular, agriculture and manufacturing sector (particularly in industrial sector) have experienced average negative growth (-4.40 per cent and -2.09 per cent per annum respectively) in between 1991 and 1996. On the other hand, recovering from the negative per annum growth in the first phase, service sector has experienced positive growth during second and third phases.

Figure 4.1

Per Annum Average Growth (%) of Population, Labour Force and Sectoral Employment

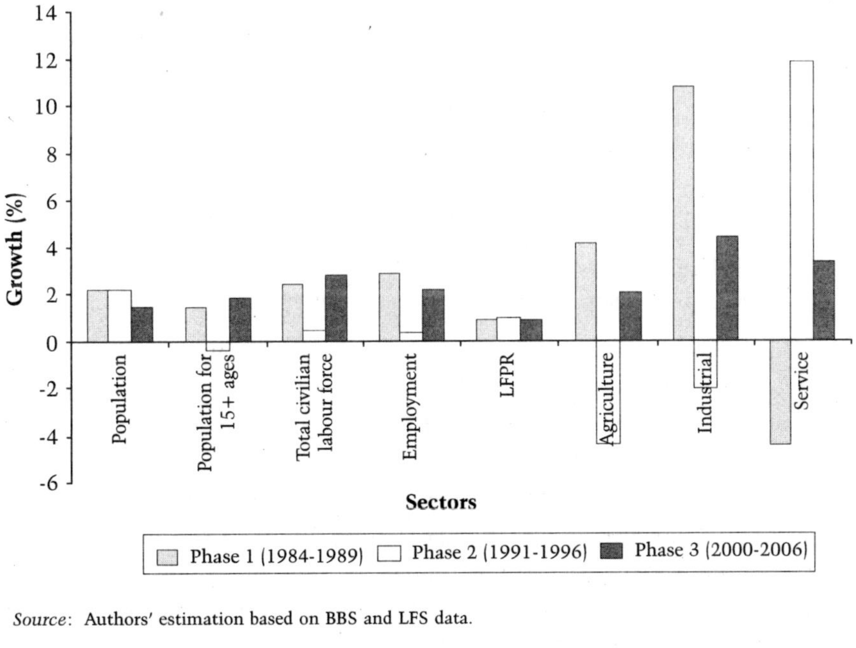

Source: Authors' estimation based on BBS and LFS data.

The dynamics of employment in overall and broad economic sectors (namely, agriculture, manufacturing and service) is shown in Annex Table A-4.1. The three sectors have different types of experiences with regard to employment trends: throughout the analysis period, manufacturing employment has remained stagnant while service sector showed an increasing trend with a big jump in 1995-96 and agriculture has experienced a decrease in 1995-96, following which employment in this sector has again increased.

It would be interesting to juxtapose the growth of sectoral contribution to GDP and the employment scenario. Theories in development economics predict that a transfer of surplus labour to the modern industrial and service sector from the traditional sector tend to take place as a country moves along its development trajectory. In the last two decades, Bangladesh has experienced some acceleration in GDP growth. Over the period of 1984-2006, the average per annum growths of real GDP and employment was 4.5 and 2.48 per cent respectively. In case of sectoral contribution to GDP and employment, substantial changes are observed for this time span.

Table 4.3 presents contribution of broad economic sectors to GDP and employment in Bangladesh over various phases of trade liberalisation. It is quite visible that the share of agriculture in GDP has been declining (coming down to 16.98 per cent of GDP in 2006 from as high as 42.7 per cent in 1984); however, this sector contributes almost half of the total employment in Bangladesh. On the other hand, increased contribution of service sector to GDP is also observed. Nonetheless, contribution of service sector to employment has experienced a decline in the late 80s but increased from mid-90s. In 2005-06, more than 40 per cent employment was generated in the services sector of the country.

The industry sector in Bangladesh is actually comprised of two sub-sectors: one is manufacturing and the other is mining and quarrying. However, contribution of mining and quarrying is negligible compared to manufacturing sector and the trend in overall industry sector, both in terms of contribution to GDP and employment, is largely dominated by industry sector. The scenario in the manufacturing sector is rather interesting. In the first phase of trade liberalisation, manufacturing sector has demonstrated a decreasing contribution, relative to the other two sectors, both in terms of value addition to the economy as well as in employment creation. From the second phase of trade liberalisation, this declining trend with regard to contribution to GDP was reversed and in 2005-06 manufacturing sector contributed about 17.1 per cent of the GDP, which was approximately 10 per cent throughout the first phase. However, the share of manufacturing employment in total labour force shows an erratic

Table 4.3

Sectoral Contribution to Real GDP and Employment Generation

(in per cent)

Sectors	*1983-84*		*1989*		*1990-91*		*1995-96*		*1999-2000*		*2002-03*		*2005-06*	
	% GDP	*% of Employment*	*% GDP*	*% of Employment*	*% GDP*	*% of Employment*	*% GDP*	*% of Employment*	*% GDP*	*% of Employment*	*% GDP*	*% of Employment*	*% GDP*	*% of Employment*
Agriculture*	42.7	58.8	38.3	65.0	37.6	66.4	28.4	48.9	24.6	50.8	22.6	51.7	17.0	48.1
Industry**	10.3	9.0	9.9	14.1	9.8	11.8	13.4	10.1	15.8	10.0	16.5	10.0	18.2	11.1
of which Manufacturing	10.1	8.9	9.8	13.9	12.9	11.8	15.4	10.1	15.4	9.5	16.0	9.8	17.1	11.0
Service	47.0	32.2	51.8	21.0	52.8	21.8	55.3	41.1	55.8	39.2	57.5	38.3	59.9	40.9
Real GDP growth rate	**4.2**		**2.5**		**3.3**		**4.6**		**5.9**		**5.3**		**6.6**	

Note: * Agriculture includes agriculture, forestry and fisheries.

**Industry includes manufacturing industries and mining and quarrying. For details of manufacturing sector please see the Annex table A-4.1.

Source: Authors' calculation based on Bangladesh Economic Review and LFS data (various issues).

pattern in the first and second phases and reached only 10.1 per cent at the end of second phase (in 1995-96) from 13.9 per cent in 1989. From the beginning of third phase, the manufacturing employment shows an increasing trend and reached 11.0 per cent in 2005-06. It is to be pointed out that share of manufacturing sector in total employment is still very low indicating the failure of manufacturing sector in absorbing the growing labour force (in a way suggested by Lewis). Possible explanation of which may include the closure and privatisation of SOEs, shrinkage of SMEs and retrenchment of excess labour in private sector for minimising costs in response to increased international competition following trade liberalisation and partly due to skill and productivity enhancement of the workers. It should also be noted here that since the early 1990s large number of Bangladeshi workers started to go abroad (mainly to Middle East as guest workers). These workers at present number about 4.1 million.

Table 4.4 shows the employment contribution of public and private sector with the formal and informal employment classification having been used for private sectors. At present, the private sector dominates the employment scenario in Bangladesh with more than 90 per cent share in the total labour force. The more interesting feature is the very strong presence of the informal sector in the private sector employment with 80 per cent share in the total labour force.[4]

Table 4.4

Employment in Formal and Informal Sector

(in per cent)

Year	*Private Sector*			*Public Sector*
	Formal Sector	*Informal Sector*	*Total*	
2002-03	16.6	79.1	95.7	4.2
1999-2000	18.6	75.3	93.9	5.9
1995-96*	8.8	87	95.8	4.2

Note: Formal sector includes individual private formal sector and private formal sector (for 2002-03).

* Based on extended definition.

Source: Compiled from various issues of LFS.

4. The recent provisional findings of LFS indicate that the share of informal and formal sector was 78.5 and 21.5 per cent respectively in 2005-06.

4.3 Status of Employment

Table 4.5 illustrates the changing pattern of labour force in different modes of employment in three phases of trade liberalisation. A major shift is observed in the composition of the status of the employment in the third phase of trade liberalisation. The self-employed category has increased to 42 per cent in 2005-06 which was about 30 per cent in the first and second phases. This increase in the self-employed category has been possible at the cost of a reduction in the unpaid family worker category which fell to 21.7 per cent in 2005-06 but was above 40 per cent in first and second phases. Indeed, till 1999-2000, one notices a decline in unpaid family workers and increases in self-employed employees as well as day labourers. But after 2000, there has been a drift back towards unpaid family workers. The share of other categories remained more or less unchanged for all the three phases. One possible hypothesis regarding self-employed population before 2000 could be that more commercial orientation of the economy, increased economic activities, thanks to increased flow of remittances and micro-credit in rural area resulted in a rise in self-employed population and a decline in unpaid family workers. However, the reverse trend after 2000 needs to be examined further to identify the underlying reasons.

Table 4.5

Status of Employment during Different Phases

(Percentage)

Category	*Phase 1*		*Phase 2*		*Phase 3*		
	1983-84	*1989-90*	*1990-91*	*1995-96*	*1999-2000**	*2002-03**	*2005-06**
Self-employed		28.57	26.49	29.3	46.7	44.8	42
Employer		1.05	0.19	0.3	0.3	0.4	0.3
Employee		9.53	11.75	12.4	16.7	13.7	13.9
Unpaid family worker		45.77	47.2	40.1	12	18.4	21.7
Domestic worker						0.6	0.7
Day labourers		15.03	13.9	17.9	24.3	20	18.2
Apprentice						0.9	0.5
Not reported (others)		0.04	0.4				2.7
Total		100	100	100	100	100	100

Note: * For 15+ and usual definition. The rest of it uses extended definition for 15+.

Source: Compiled from various issuers of LFS.

Box 4.1

A Formidable Macroeconomic Challenge for Attaining PRSP Target and Beyond: Urgent Need for Comprehensive Employment Strategy[5]

Table: Actual Employment Scenario *vis-à-vis* PRSP Projection (In Million)

	FY 03 (LFS)	*FY 06 (LFS)*	*FY 06 (PRSP)*	*Shortfall in Employment*	*FY 07 (PRSP)*	*FY 08 (PRSP)*
National	44.3	47.4	51.69	(-) 4.29	54.75	58.08
Rural	33.6	36.1	38.3	(-) 2.20	40.35	42.57
Urban	10.7	11.3	13.39	(-) 2.09	14.4	15.51

Source: Authors' estimation based on PRSP (Table 7) and LFS reports (various reports).

According to the PRSP, the projected total employment for FY 2006 was 51.69 million against a base of 44.03 million in FY 2003 (from LFS 2003). That would imply a compound annual growth rate (CAGR) of 5.49 per cent during this period. However, analysis of LFS data (2003 and 2006) reveals that the compound annual growth rate (CAGR) of actual overall employment from FY 03 to FY 06 was 2.28 per cent, with the rural rate being 2.42 per cent and the urban rate being 1.84 per cent. The PRSP document has also set the target of total employment to be 58.08 million in FY08 against the base of FY 03 implying a CAGR of 5.70 per cent between FY 2003 and FY 2008. Given the present employment scenario as reported in LFS of FY 06, to achieve this projected employment size in FY 2008, overall employment has to be generated at the rate of 10.69 per cent, with rural being by 8.59 per cent and urban being by 17.16 per cent annually over the next two years (if the actual employment figure for 2006 is taken into account). Given the current macroeconomic scenario and the recent employment and growth relationship experience, this will indeed be a Herculean task. Considering the recent trends of urbanisation in the country and slow pace of employment creation in urban areas, achieving the PRSP employment targets will be an issue of critical importance from the perspective of policy-making.

contd...

5. In other studies also, the magnitude of the employment challenge has appeared to be quite serious for Bangladesh. Islam (2007) mentioned that, "With the observed employment elasticity of 0.495, an employment growth of 4.4 per cent (i.e., the same rate as labour force growth, not counting the currently unemployed and underemployed) would require a GDP growth of 8.89 per cent per annum. If the employment elasticity is assumed to be 0.39 (as estimated by ILO, 2002 for the period 1990-91 to 1995-96), the required GDP growth would naturally be higher."

...contd...

Figure: Achieved Annual Growth *versus* Required Annual Growth to Attain PRSP Target

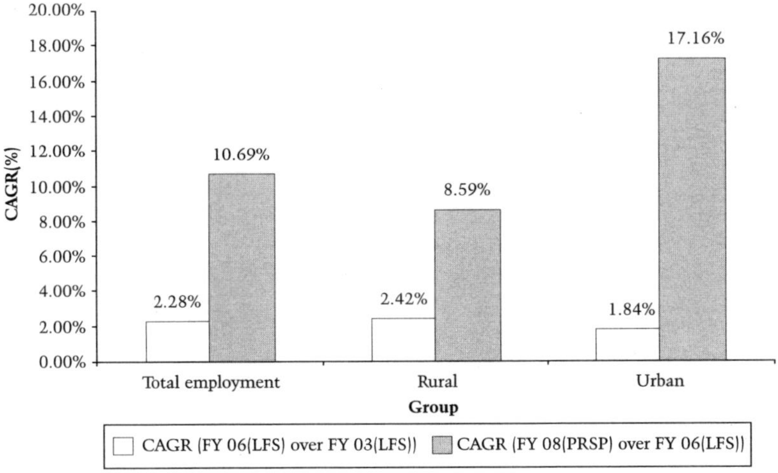

4.4 Unemployment and Underemployment

As revealed by the LFS data, there was an increase in total employment over the past years; on the other hand, unemployment rate was also increasing, at a higher rate, indicating that the labour force was growing at a faster rate compared to the rate of creation of new job opportunities. Table 4.6 shows the unemployment rate in Bangladesh during various phases of trade liberalisation.

Table 4.6 illustrates that total unemployment rate experiences a rise in all the phases. In general, urban unemployment rate is found to be higher than the rural while unemployment is more prevalent among women compared to their male counterpart. Unemployment rate varies between one and six per cent for the period under consideration. However, these figures do not match with the reality and are much lower compared to even the rates usually prevailing in the developed countries. This is mainly due to definitional problem of unemployment in LFS reports which are based in the context of industrial economies. LFS definition contains 'Not employed during the last two weeks and willing to work at the current wage rate' which is suitable for our formal sector but is not applicable for the large informal labour market that is quite prominent in Bangladesh.

Table 4.6

Unemployment Rate by Sex and Residence (By Usual Definition)

(Percentage)

	Phase 1		*Phase 2*		*Phase 3*		
	1983-84	*1989*	*1990-91*	*1995-96*	*1999-2000*	*2002-03*	*2005-06*
Bangladesh							
Total	1.8	1.2	1.9	2.5	4.3	4.3	4.2
Male	1.5	1.3	2	2.7	3.4	4.2	3.4
Female	4.0	1.0	1.9	2.3	7.8	3.5	7.0
Urban							
Total	2.8	3.5	2.6	4.3	5.8	5	n/a
Male	2.6	2.4	2	4.3	5	4.6	n/a
Female	3.9	6.7	4.7	4.2	8.3	6.2	n/a
Rural							
Total	1.6	0.9	1.8	2.1	3.9	4.1	n/a
Male	1.4	1.2	2	2.2	2.9	4	n/a
Female	4.2	0.5	1.6	1.9	7.6	4.4	n/a

Note: Unemployed person is a person who is involuntarily out of gainful employment during the reference period but either: (a) has been actively looking for a job or (b) was willing to work but not looking for work because of illness or believing that no work was available.

Source: Authors' estimation based on BBS and LFS data.

Nevertheless, one can still observe that from the second phase of trade liberalisation the unemployment rate has been increasing more rapidly in all cross-sections of the population.

Table 4.7 illustrates the underemployment rate of the country which, in view of the shortcomings in the definition of unemployment, is a better indicator of evidence as regards the surplus labour. Though the total underemployment rate is on a decline, more than one-third of the labour force continues to be underemployed. The underemployment rate is lower in urban areas as compared to rural areas where informal employment is widespread and sharing of household's consumption basket continues to be the tradition. However, it is to be noted that there is significantly higher underemployment rate for females compared to males. Therefore, in terms of underemployment the rural females are worse off, whereas urban males are most extensively employed in the labour market.

Table 4.7

Underemployment Rates in Different Phases

(Percentage)

	Underemployment Rates of Population by Sex and Locality						
	Phase 1		Phase 2		Phase 3		
Under-employment Rate(%)	*1983-84*	*1989*	*1990-91*	*1995-96**	*1999-2000**	*2002-03***	*2005-06***
Bangladesh							
Total		43.4	42.8	38.5	31.9	34.2	24.5
Male		15.4	15.7	13.7	8.4	23.1	10.9
Female		83.0	85.3	79.0	71.2	72.3	68.3
Urban							
Total		21.6	29.8	22.1	18.2	26.7	
Male		6.7	18.9	10.9	4.9	16.0	
Female		63.6	65.6	54.2	52.0	71.8	
Rural							
Total		46.1	45.6	42.1	35.0	36.4	
Male		16.8	14.8	14.5	9.3	25.3	
Female		84.4	87.2	82.4	74.2	75.9	

Note: Underemployment is defined as percentage of those who worked less than 35 hours per week as a percentage of total number of employed person.

* based on extended definition 15+ age, **based on usual definition and 15+ age. The rest are for 10+ age and usual definition.

Source: Authors' estimation based on BBS and LFS data.

Though open unemployment and underemployment reflect two different phenomena, there are apprehensions that some of the underemployed workers could be misclassified as underemployed and vice versa. Indeed, it is interesting to note that while open unemployment has risen, the degree of underemployment (by time measure) has fallen. While a full explanation of these phenomena is beyond the scope of the present chapter, proper explanation for these opposing trends merits further in-depth study. Along with the analyses on gender dimension and rural-urban classification, the age structure of the unemployed and underemployed population could also provide useful insights about labour market dynamics. However, it is to be noted that the structural changes that are taking place in the economy, especially in the rural segment of the economy, may have some implications in this regard. As mentioned earlier, commercial orientation of the economy in the recent years could have resulted in more open unemployment. On the other hand, given the increase in economic activity (especially in rural areas due to a variety of programmes (including micro credit related), it is possible that people are spending more time in these activities compared to the earlier period leading to a lowering of underemployment (by a pure time measure).

4.5 Gender Dimension of Employment: Female Participation Rate (FPR)

As in many South Asian and other developing countries, female participation rate (FPR) in economic activities is relatively low in Bangladesh. Women are mainly engaged in low income and peripheral jobs. The following table shows female participation rates over the various phases of trade liberalisation.

Table 4.8

Female Participation Rate

(in percentage)

	Phase 1		*Phase 2*		*Phase 3*		
In %	*1983-84*	*1988-89*	*1990-91*	*1995-96*	*1999-2000*	*2002-03*	*2005-06*
Agriculture	1.4	45.5	50.3	47.8	18.7	25.1	Na
Industry	24	63.6	28.4	36.8	37.7	38.6	Na
Service	20.4	15.8	13.4	21.2	22.14	16.2	Na
Total	8.4	41.4	39.3	38	20.2 (23.9)	22.2 (26.1)	29.2

Note: From 1989, LFS report some particular dataset quality is ambiguous because of definitional problems.

Source: Compiled from various issues of LFS.

If the data is to be believed, the overall female participation rate had experienced a major hike between 1983 and 1989, a rise to 41.4 per cent from a very low base rate of 8.4 in 1983. The rate remained stable during second phase which was followed by a decrease in the third phase when it dropped to below 30 per cent. The key reason behind the extraordinary growth in 1989 over the 1984 period is mainly explained by definitions' changes. Since 1989, LFS data has included a number of additional activities such as ploughing, irrigation, harvesting, collecting, cleaning, drying or boiling, growing vegetables or spices, poultry raising etc. These activities are generally performed by females which in turn increased total female labour force participation rates. Since most of the newly included activities in 1989 are agriculture-centric, for quite obvious reasons FPR in agricultural sector showed a similar pattern as in the overall FPR: sudden rise between 1984 and 1989 then stabilisation in second phase followed by a decline.

Similar to agriculture, FPR for industry experienced a sharp rise in the phase one with an abrupt jump in FPR of 40 percentage point (from 24 per cent to 64 per cent) between 1984 and 1989. However, surprisingly, the rate had dropped by 35 percentage point as is reported in the immediate next LFS of 1990-91 and stayed stable to the tune of 37-38 per cent over the remaining time period. The hike in first phase may be explained due to the expansion of the export-oriented RMG sector in Bangladesh. However, the drastic fall in FPR in the next LFS report of 1990-91 cannot be explained properly. On the other hand, service sector experienced a more erratic movement: from 20.4 per cent in 1983-84, the FPR declined 7 percentage points in 1990-91 followed by an increase of 8 percentage points in 1995-96 which again declined to 16.2 per cent in 2002-03.

In sum, no conclusive evidence can be found as regards the impact of trade liberalisation on female participation in labour force. If the latest LFS data are taken to be more reliable, it can merely be said that overall the current female participation rate is about 29 per cent in Bangladesh.

4.6 Real Wage Index and Labour Productivity Index at Aggregate Level

According to Stolper-Samuelson theorem, trade liberalisation increases real returns of sectors which use abundant factors of production more intensively. As a labour abundant country, it thus would be expected (according to S-S theorem) that liberalisation of trade will increase real wage of labour. It is from this perspective, the real wage index[6] was estimated to look at the movement of wages over the past years.

6. Nominal wage index was corrected for changes in purchasing power measured by the CPI. For Bangladesh, the real wage index had an earlier series with base year of 1969-70, while a new series with base year 1989-90 is also available. In this analysis, the entire real wage index has been rebased to 1989-90.

Figure 4.2

Real Wage Indices by Major Sectors (Base: 1989-90=100)

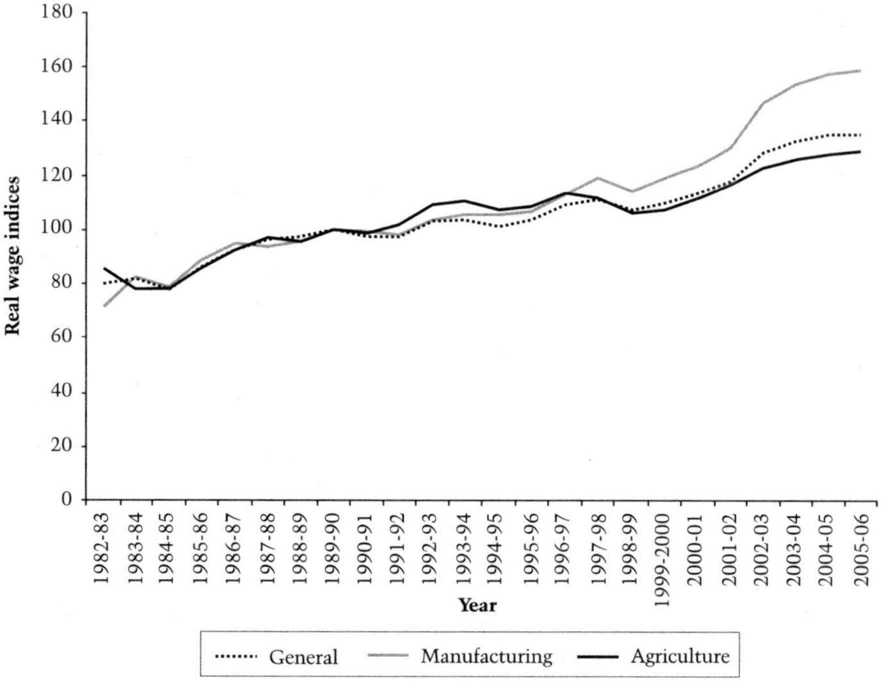

Source: SYB, BBS and Sources and Methods of Indices (National accounts Deflators) (published by Statistical Year Book, BBS).

Real wage indices for phase one implies increasing trend of wage indices for all three categories: general, manufacturing and agricultural sector. For all these sectors, an upward movement is observed. Though the nominal wage increased during the initial period of phase two (rapid trade liberalisation), real wages for general and manufacturing sector tended to decline (mainly in 1991 and 1992). However, from 1993 this declining trend reversed and all sectors have demonstrated a rising trend. From phase three, rapid upward movement of wages in manufacturing sector (except 1998-99) is observed with fluctuations for general and agricultural sectors.

In developed countries, a positive correlation generally tends to exist between real wages and labour productivity. In other words, real wages and

Table 4.9

Labour Productivity Index (and Real Monetary Value in '000 Taka) for Major Economic Sectors (Base Year: 1990-91=100)

Year	Agriculture	Manufacturing	Service	Overall
1983-84	176.67 (19.97)	180.66 (54.39)	79.15 (53.18)	127.80 (33.8)
1988-89	91.45 (10.33)	69.05 (20.79)	84.96 (57.09)	81.77 (21.6)
1990-91	100.00 (11.30)	100.00 (30.11)	100.00 (67.19)	100 (26.4)
1995-96	168.93 (19.09)	231.40 (69.67)	96.51 (64.85)	181.02 (47.8)
1999-2000	225.52 (25.49)	271.07 (81.61)	111.22 (74.73)	199.82 (52.8)
2002-03	206.87 (23.38)	278.99 (84.00)	119.17 (80.07)	202.78 (53.6)
2005-06	180.63 (20.41)	299.05 (90.04)	126.56 (85.04)	227.72 (60.2)

Note: Figures in the parenthesis indicate labour productivity per employee per year in 000 taka. For a particular sector and for a specific year, labour productivity is estimated by dividing the contribution of that sector to real GDP by the number of employee engaged in that sector.

Source: Authors' estimation based on various issues of LFS (for employment data), WDI (for GDP data for 1984 and 1989) and BBS (for 1990-91 to 2005-06 GDP data).

productivity grow at the same rates in the long run. However, in the short run this relationship can be affected by various market agents e.g., the government, trade unions, powerful business lobbies etc. Sometimes, at the aggregate level this relationship is found to be true, but this is not necessarily true at the sectoral level (Chubrik *et al.*, 2006). On the other hand, in case of developing countries, empirical examination could reveal absence of such relationship. For example, Kirmanoglu and Yazgan (1999) showed lack of co-movements in wage and productivity in Turkish manufacturing factories. In addition, for the public sector, they found inelastic response of real wage to labour productivity. Several studies reveal stagnant real wage (with increasing labour productivity) over time due to forces of globalisation, greater capital mobility and the increasing international opportunities in labour market.

Table 4.9 and Figures 4.3-4.5 show labour productivity indices for overall economy as well as for the three major economic sectors for seven years covering between 1983-84 and 2005-06.[7] Since the data on real wages for service sector is not available for Bangladesh, the movements of labour productivity for service sector could not be compared with corresponding wage index.

It can be seen that following the 36 per cent fall in labour productivity index for the overall economy, between 1983-84 and 1988-89, the index shows persistently positive trend to reach 227.7 in 2005-06 which is equivalent to 60.2 thousand taka per employed person per year in real terms. Since no drastic fall in the labour input was observed between 1984 and 1989, one can identify the fall in output due to the devastating floods in consecutive years of 1987 and 1988 as the key reasons behind the fall in labour productivity in 1988-89. It is to be noted that the highest increase in the index is observed in 1995-96. On the other hand, the real wage index for general category is quite stable and positive growth is observed throughout the years under consideration.

Figure 4.3

Dynamics of Labour Productivity and Real Wage for Overall Economy

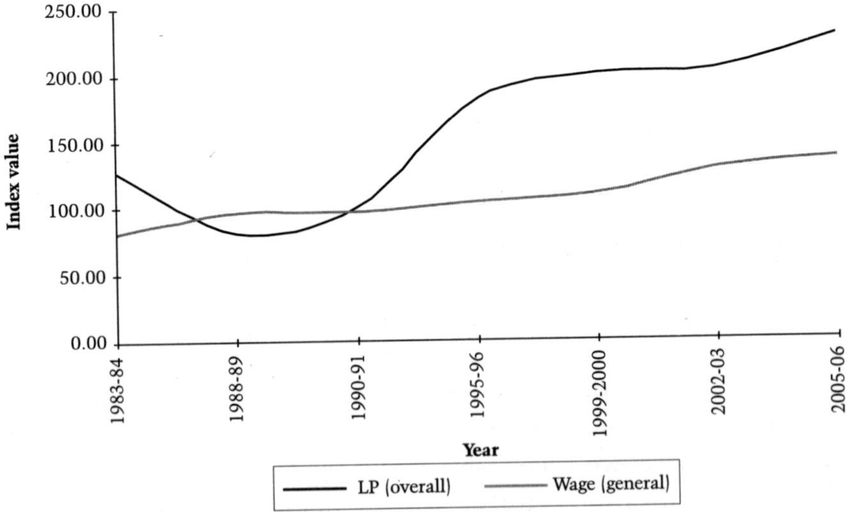

Source: Authors' estimation based on BBS and LFS various issues.

7. Though yearly GDP data is available for the entire period, employment data is only available from the LFS reports. Therefore, labour productivity for seven years (for which both GDP and employment data is available) can be estimated.

If the broad economic sectors are examined, similar pattern for labour productivity index is observed for manufacturing sector with a decrease in 1988-89 and upward trends during the remaining period. However, steady and positive growth of the real wage index is observed. Interestingly, labour productivity has almost doubled during 1995-96 whereas a lagged reflection of this jump is partially observed for real wage indices in 1999-2000 and 2002-03 (Figure 4.4). Although more investigation is needed to answer this puzzle relating to inferior response of real wage to productivity rise, one possible explanation could be the relatively weak bargaining power of labour *vis-à-vis* employers in an environment of weak trade union movement and surplus labour.

Figure 4.4

Dynamics of Labour Productivity and Real Wage for Manufacturing Sector

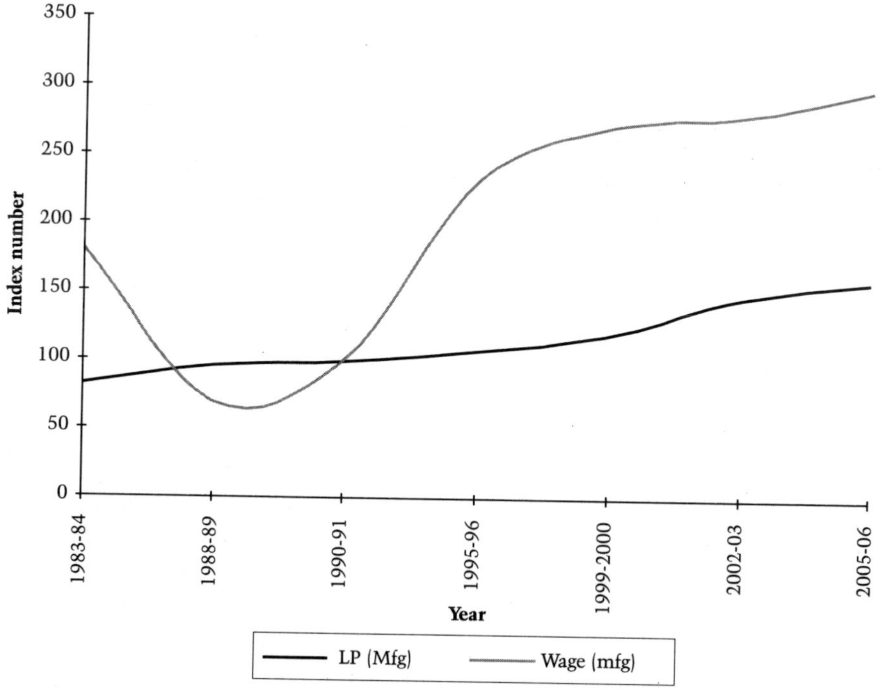

Source: Authors' estimation based on BBS and LFS various issues.

Labour productivity of agricultural sector displays analogous pattern with a decrease in 1988-89 followed by an upward trend till 1999-2000 when the index reached 225. However, following this, the productivity trends have reversed and come down to 180 in 2005-06. It can be argued that given the relative factor endowments (i.e., land, labour, technological status etc.) in the economy of Bangladesh if a sector (in this case agriculture) approaches the maximum production frontier, productivity could remain stagnant, slow down, and even in some cases could come down. This fall in productivity may not, however, necessarily lead to a fall in real agricultural wage index. Rather, in case of Bangladesh the real wage of agricultural sector continued to grow and reached 129 (in 2005-06) from 107 (in 1999-2000).

Figure 4.5

Dynamics of Labour Productivity and Real Wage for Agricultural Sector

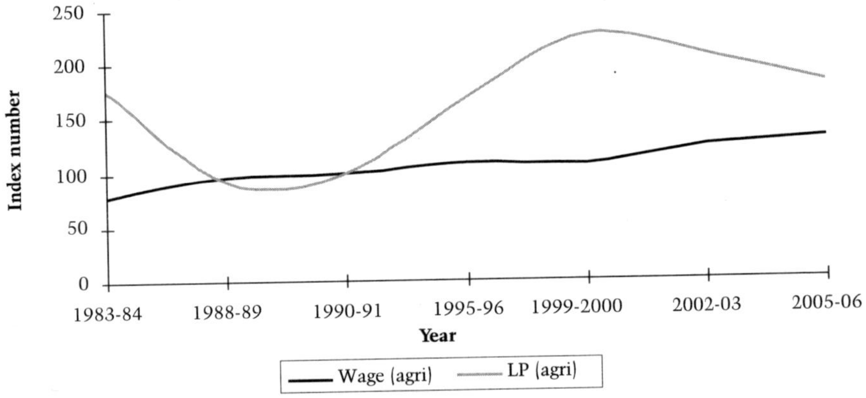

Source: Authors' estimation based on BBS and LFS various issues.

Table 4.10 displays the bivariate correlation coefficient between wage and productivity indices for the overall economy, manufacturing sector and the agricultural sector. Though this correlation analysis does not imply any causal relationship between the indices, it could provide important insights as regards the co-movements of wage and productivity in the economy. It is evident that all wage and labour productivity indices, except labour productivity for agriculture, show significant, positive and strong correlation with time indicating the rise in productivity as well as wages over the time. Significant positive correlations between real wage indices (between labour productivity indices) of different sectors signify the co-movements of wage (labour productivity) of various sectors of the economy.

The most interesting aspect of this analysis lies in the correlations between the wage and productivity indices of a particular sector. One can see that labour productivity of overall and manufacturing sector have significant and strong positive correlations with corresponding real wages—the respective coefficient values are 0.800 and 0.763. However, as was mentioned earlier, in case of agriculture, no significant association between productivity and wages could be found. As mentioned earlier, lack of wage data restricts an examination of this relationship for the service sector. Considering the dominance of services in the overall economy (see Table 4.3), one could justifiably discern a positive alliance between wage and productivity in case of service sector as well.

Table 4.10

Correlations Matrix between Labour Productivity (LP) and Real Wage Indices

	Year	*Wage (General)*	*LP (Overall)*	*Wage (Mfg.)*	*LP (Mfg.)*	*Wage (Agri)*	*LP (Agri)*
Year	1						
Wage (general)	0.969*	1					
LP (overall)	0.876*	**0.800****	1				
Wage (mfg.)	0.957*	0.993*	0.825**	1			
LP (mfg.)	0.803**	0.725	0.988*	**0.763****	1		
Wage (agri)	0.969*	0.982*	0.77*	0.957*	0.678	1	
LP (agri)	0.572	0.457	0.849*	0.507	0.912*	**0.385**	1

Note: Number of Observations: 7

* and ** denote correlation is significant at the 0.01 level and 0.05 level respectively (2-tailed).

Source: Authors' estimation based on BBS and LFS various issues.

Although evidence is found favouring a linear association between the real wage and productivity for the overall economy, as also for the manufacturing and service sectors, and this can be reasonably explained, the growth in real wages lagged behind that of productivity. It is clearly visible from the study that these two indices diverge over time for both overall and manufacturing sectors and the gap between these is highest in case of the manufacturing sector.

These findings lead to the following plausible hypotheses which are, of course, subject to empirical verification. The sectoral real wage elasticities with respect to corresponding labour productivity would perhaps vary from sector to sector. In case of agriculture, the relative movements in real wage and productivity would suggest inelastic behaviour of agricultural real wages with respect to productivity rise. If productivity in agricultural sector increases and if wage does not respond positively to the increase in productivity, then it would imply that workers are not receiving enough gains from increased productivity. If the reverse happens, i.e., if wage remains sticky or increases when the productivity falls, it implies that the workers are getting protection against fall in productivity. In case of manufacturing sector the elasticity, although positive, would seem to be low compared to that of the overall economy. If wage remains sticky during the period of rising productivity in manufacturing sector, this would indicate that workers are not getting a share in the rising productivity through rising real wages. If the reverse happens in the manufacturing sector, i.e., if wage remains sticky or increases during the period of decreasing productivity, this may lead to a trade-off between real wage determination and fall in competitiveness.[8]

8. However, without proper econometric analysis and data availability, this hypothesis cannot be tested.

Annex A-4.1

Notes on the Definition of Indicators of LFS

1. Labour force

Economically, active population or labour force is defined as persons aged 15 years and above, who are either employed or unemployed during the reference period of the survey.

2. Labour force participation rate

It is defined as total civilian labour force divided by total population of 15+ ages.

3. Unemployed person

Unemployed person is a person who is involuntarily out of gainful employment but either:

a. has been actively looking for a job or

b. was willing to work but not looking for work because of illness or believing that no work is available.

4. Underemployment

It is defined as percentage of those who worked less than 35 hours per week as a percentage of total number of employed person.

5. Female participation rate

It is calculated as a percentage of female employment to total employment.

Annex Table A-4.1

Sectoral Distribution of Employment (in Million)

	Phase 1		*Phase 2*		*Phase 3*		
	1983-84	*1989*	*1990-91*	*1995-96*	*2000*	*2002-03*	*2005-06*
Employment	27.11	31.62	33.55	34.80	38.98	44.32	47.70
Agriculture	15.94	20.54	22.28	17.00	19.79	22.93	22.80
Mining	0.05	0.06	0.01	0.00	0.17	0.08	0.10
Manufacturing	2.41	4.40	3.96	3.50	3.72	4.34	5.20
Total manufacturing (mining+manufacturing)	2.45	4.46	3.97	3.50	3.90	4.43	5.30
a. Electricity, gas, water	0.07	0.01	0.03	0.10	0.13	0.10	0.10
b. Construction	0.47	0.42	0.35	1.00	1.10	1.54	1.50
c. Trade and restaurant	3.15	2.60	2.87	6.00	6.15	6.67	7.80
d. Transport and communication	1.05	0.81	1.08	2.20	2.47	3.02	4.00
e. Finance and business service	0.13	0.15	0.20	0.20	0.40	0.22	0.70
f. Community and personal service	2.22	1.13	1.28	4.80	5.04	2.55	2.60
g. Others	1.62	1.51	1.50	0.00	0.00	2.87	2.60
Total service (a+b+c+ d+e+f+g)	8.72	6.63	7.30	14.30	15.30	16.97	19.30

Source: Various issues of LFS, BBS.

5 An Analysis of Employment Situation in the Manufacturing Sector

This chapter will analyse phase-wise changes in the industrial employment situation along with key industrial performance indicators based on an analysis of the CMI data. The trends in labour productivity and real wages have also been examined. The analysis has been conducted for overall industrial sector; analysis has also been carried out for public *versus* private, export oriented *versus* non-export oriented, and labour intensive *vis-à-vis* non-labour intensive sectors to explore and examine any significant inherent dynamics in these sub-categories of economic activities with respect to trade reforms and employment implications. These analyses of broad categories supplemented by analyses of major industrial sectors at 4-digit level disaggregate data. Prior to investigating the manufacturing employment situation of Bangladesh, based on the CMI data, the relationship between protection and industrial growth performance has been examined since expansion of industrial capacity automatically leads to higher employment

5.1 Trade Reforms, ERP and Production Performance

The effective rate of protection (ERP) takes into account the total effect of tariffs on the inputs and outputs in a production process and provides a picture of the real protection enjoyed by certain economic activities. One of the salient features of trade reforms tend to be lowering of the anti-export bias (along with reduction in ERP) to stimulate export intensive industrial growth. Two sector-wise ERP data series are available, both prepared at Bangladesh Tariff Commission, providing sector-wise ERPs.[1] However, there are some data inconsistencies between these two

1. In old series, the time period covered is 1992-93 to 1999-2000 while the new series covers 1993-94 to 2002-03. However, it is to be mentioned here that the new ERP series is yet to publish publicly and subject to revision.

series. For example, old ERP series yields negative growths for cement, leather and fish and fish products over the initial and end year, while positive growths are observed in case of the new series. Even in few cases where both the series indicate declining ERPs, the percentage changes of ERP shows considerable divergence e.g., for pharmaceuticals industries, cloth mill, tobacco etc. (Table 5.1).

Table 5.1

Growth of Old and New ERP with Quantum Growth of Industries

BSIC Code	*Industry Description*	*Growth of Old ERPs (FY 2000 over FY 1993)*	*% Change in Production (FY 2000 over FY 1993)*	*Growth of New ERPs (FY 2003 over FY 1994)*	*% Change in Production (FY 2003 over FY 1994)*
Import Substituting and Import Competing Sectors					
3123	Sugar	-67.71	-34.17	Na	Na
3211	Yarn	-53.2	6.41	-66.2	37.98
3115	Edible Oil	-53.21	-38.67	-59.38	-1.5
3411	Paper	-77.31	-38.75	-77.71	-66.55
3511	Pharmaceutical	-213.33	48.71	-77.89	115.78
3692	Cement	-62.14	562.23	146.31	442.75
3141	Tobacco cigarette	-48.73	71.35	-141.05	77.78
3211	Cloth mill	-61.68	-72.5	-141.05	-42.85
3524	Chemical fertiliser	-110.71	55.22	-121.47	-4.34
37	Iron and steel	-52.32	174.16	-31.53	86.5
38	Metal products	-70.78	8.64	-49.52	22.83
383	Machinery	-89.05	-60.38	-62.73	12.45
385	Transport equipment	-74.39	-13.71	-8.12	-12.3
Export Oriented Sectors					
	Jute	-58.38	-20.3	-50.988	-1.01
3126	Tea	-50.42	4.26	-57.07	5.56
3114	Fish and fish products	-70.41	48.67	14.03	16.07
3231	RMG	-75.17	167.07	-70.18	194.25
3213	Jute textile	-55.7	-23.38	-10.69	-24.11
324	Leather and leather products	-93.41	42.03	39.35	4.21

Source: Authors' estimation based on Bangladesh Tariff Commission and BBS.

It has been argued that reduction in ERPs enhances the industrial production capacity particularly of the export oriented sectors. To investigate the impact of ERP on the growth of industries (as measured by quantum index) over the past years in Bangladesh, the changes in ERP and quantum index are juxtaposed in Table 5.1.[2] Industries are divided into two major groups for this purpose: export-oriented sectors, and import substituting and import-competing sectors.

(a) Export-oriented Sectors: In case of six export industries, according to the old series, ERPs have declined for all sectors. On the other hand, for the set of new ERPs, a rise in the protection for fish and fish products and leather and leather products is visible. While looking at the growth performance, it can be observed that only jute and jute textiles has experienced a fall in production.

(b) Import-substituting and Import-competing Sectors: It is found from both ERP estimates that in case of import-substituting industries in Bangladesh, ERP has decreased for all sectors except cement. In case of cement, old ERP shows a decline whereas new ERP shows an increase. Nonetheless, during both periods, covered by these two series, productions of these sectors have experienced significant increase. For both old and new ERP series, out of the remaining 12 import-substituting and competing sectors, negative growth performance is observed for five sectors over the corresponding period indicating adverse impact of decreasing protection as a result of trade reforms.[3] However, a number of sectors including yarn, pharmaceuticals, tobacco, cigarette, chemical fertiliser, iron, steel and metal products have posted positive growth during the trade reform period.

Thus, whilst there has been a decline in the protection rate in general, as a consequence of trade reforms, one finds both winners and losers (in terms of growth performance) among both export-oriented and import substitution industries in Bangladesh.

---

2. See Annex Tables A-5.1 and A-5.2 for further details.
3. These sectors are: sugar, edible oil, paper, cloth mill and transport equipment for old ERP while edible oil, paper, cloth mill, chemical fertiliser and transport equipment for new ERP.

Growth performance of various industries in Bangladesh does not show any unilinear relationship with movements in the ERP. Rather, a mixed picture is observed during the period of trade reform and decline in protection. Although some of the sectors have been able to demonstrate higher growth rate during trade reform period, some others have suffered. It is important to recognise that other important factors besides ERP tend to play an important role in determining growth performance including business support services, adequate infrastructure, availability of skilled labour, access to credit on reasonable terms and good governance.

5.2 Trade Reforms, Manufacturing Employment and Output

CMI data reveals that manufacturing employment both in overall production and also non-production related employment have, in general, increased during different phases of trade liberalisation as well as in pre-reform period.[4] Figure 5.1 shows the growth rate of production workers (PW), non-production worker (NPW) and number of employees (NOE) from 1973 to 2002.[5] It is to be noted that PW constitutes the major portion of the NOE so that movement in PW growth is more likely to be reflected in the growth of NOE. Negative growth in PW is revealed for 1976, 1977, 1983 and 1994 which in turn induce negative growth in NOE. On the other hand, negative growth for NPW is observed for 1983, 1986, 1987, 1991, 1993, 1994 and 2000; out of these seven years, for 1993 the strong negative growth (- 47.5 per cent) of NPW offsets the positive growth in PW resulting in a negative growth of -5.9 per cent for NOE. One very interesting observation found from Figure 5.1 is the hike in growth rate for 1989 for all types of employees (See Box 5.1). It is also to be noted that despite a trivial growth of 4 per cent in PW, an unusual growth of 103 per cent was observed in 1992 for NPW resulting in a formidable growth of about 20 per cent of NOE.

4. Total manufacturing employment is the sum of production and non-production worker (reported as total person engaged in the CMI), while number of employees equals production worker and administrative workers.

5. A brief definitional note on various CMI terminologies used in this study is presented in Annex A-5.1.

Figure 5.1

Dynamics of Growth Rate of PW, NPW and NOE (From 1973 to 2002)

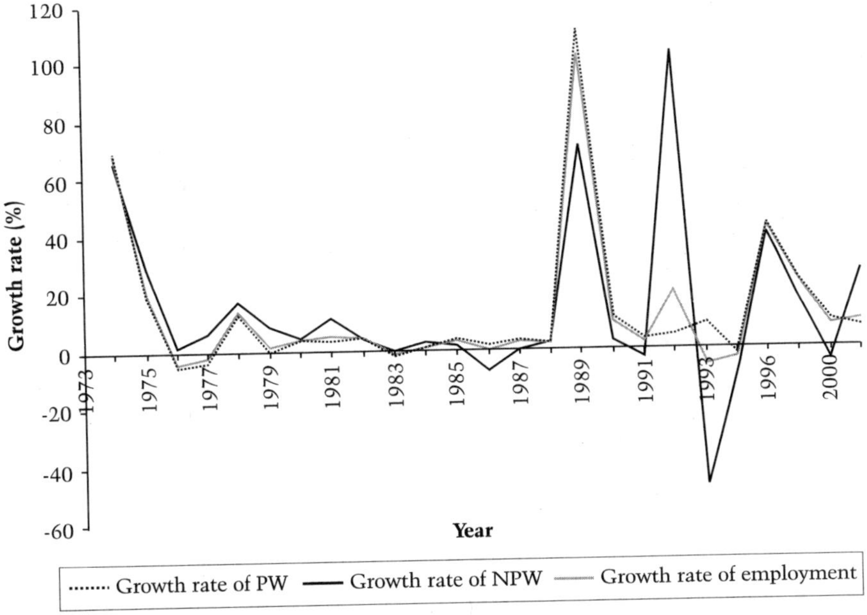

Source: Authors' estimation based on various issues of CMI.

The per annum growth rates for the period 1973-2002 was 9.1, 9.2 and 8.6 per cent for NOE, PW and NPW respectively. If the phase-wise per annum growth rates are compared, it can been seen that the pre-reform period experienced the highest growth of manufacturing employment, in general, followed by first phase and third phase (Figure 5.2). Most importantly, the growth for PW during the first phase of reform period was the highest, though the growth in NPW was relatively low in this phase compared to pre-reform period. As a matter of fact, in none of the reform phases NPW experienced as high a growth as was the case for the pre-reform period. This may indicate that because of the increased competition induced by opening up of the economy, manufacturing sector shifted more towards growth of the production workers and their productivity. However, before considering the impressive growth performance of the first phase, the structural change in data during 1989 (See Box 5.1) should be kept in

Box 5.1

1989: A Structural Break in CMI Data Series

For 1989, the CMI data shows significant increase in all industrial statistics e.g., number of establishment, all employee (with operational workers), gross output etc. In addition, growth in this year indicates substantial increase in growth of employees, productive and non-productive workers. For example, in case of growth rate of number of employees, production workers and non-production workers, the rate was 2.26, 2.36 and 1.87 per cent in 1988 respectively that increased exorbitantly to 102.59, 110.46 and 71.47 per cent in 1989. Again after 1989, all statistical indicators behave quite smoothly which implies that 1989 should be considered as a structural break in the industrial data series in Bangladesh. The key reason for such an increase is that in 1988-89 CMI, steps were taken to cover both the industrial units registered with the CIF (Chief Inspector of Factories) as well as unregistered (at that point of time) to produce more comprehensive manufacturing statistics. Therefore, number of establishments increased from 4793 in 1988 to 23752 in 1989 (with inclusion of large number of handloom establishments in Rajshahi Division).

Incremental Contribution of Major Sectors in Increased Coverage and Employment during 1989 CMI

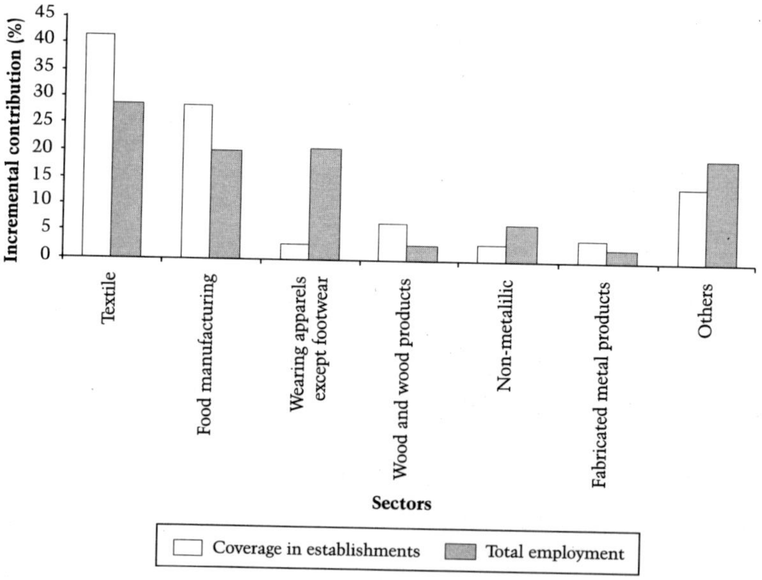

It is to be noted that major expansion in coverage is observed for the private sector. Major increments between 1988 and 1989 is found for textile sector (from 2316 to 10153), food manufacturing (from 635 to 6001), wearing apparels except footwear (from 84 to 639), wood and wood products (from 32 to 1293), non-metallic production (from 21 to 571), fabricated metal products (from 332 to 1067). Most of these sectors are labour-intensive and export-oriented indicating huge increase in growth rate of productive workers in 1989 (See Figures 5.7 and 5.8). As a result, growth rate of private sector employees in 1989 was 168.14 per cent while for public sector, it was 24.25 per cent.

mind. The second phase of trade liberalisation experienced a decline in employment growth compared to all other phases, possibly an outcome of the adjustment process through which the industrial sector passed during the reform period. Though NOE and PW have still experienced positive growth during the second phase, albeit to a lower extent (2.0 and 2.5 per cent, respectively), negative average growth of NPW (-1.0 per cent) is also observed for these period. However, the third phase of the liberalisation appears to have recovered from the adverse impact of adjustment costs.

Figure 5.2

Phase-wise Comparison of Per Annum Growth (%) of Industrial Employment

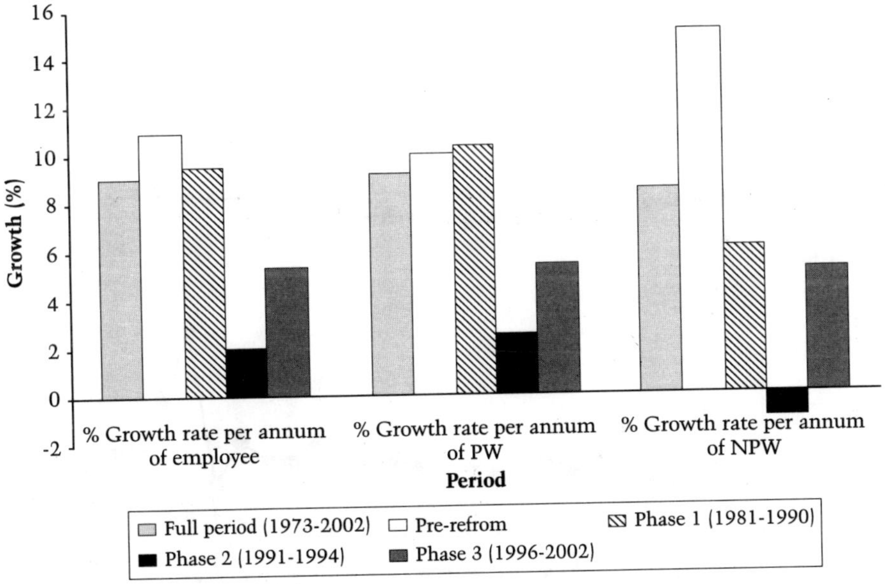

Source: Authors' estimation based on CMI data, various issues.

Another key and interesting feature of this analysis was an examination of the impact of trade liberalisation on production related employment. Figure 5.3 shows trends in PW with its share in NOE and growth rates over 1973-2002. It is evident from Figure 5.3 that production workers constitute more than three-fourth of the total number of employees; more precisely, the shares of PW lie between 74 to 89 per cent over the time period under examination. At the end of the second phase of trade liberalisation, there were around one million production worker which doubled in the span of just eight years to cross the two million mark in 2002. Hence, it can be said that during the period of trade liberalisation manufacturing employment has increased in absolute term; however, how much of this increase is solely due to trade reforms controlling for other factors and what is the likely impact of further liberalisation is a critical question that merits consideration. Chapters 6 and 7 will try to answer these questions.

Figure 5.3

Trend in Total Number of Production Workers, its Growth and Share in Total Number of Employee

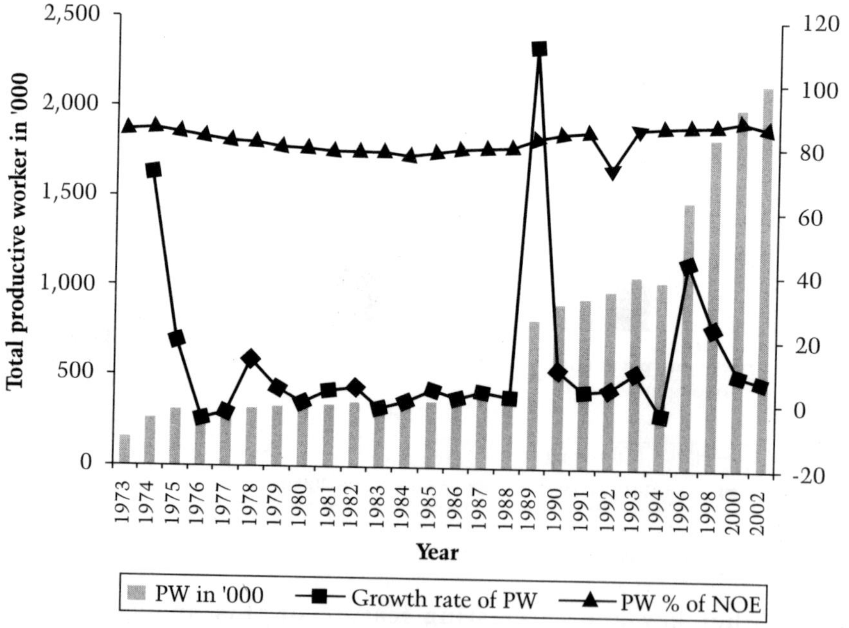

Source: Authors' estimation based on CMI data, various issues.

Box 5.2

Mismatch between Industrial Employment Growth in LFS and CMI

Figure: Growth Rates (%) in Manufacturing Employment: LFS *Vis-a-vis* CMI

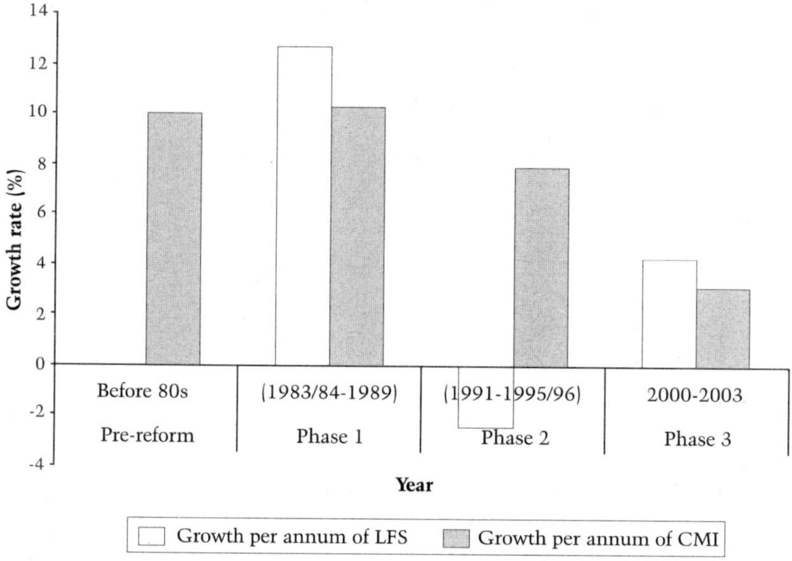

Source: Various issues of LFS and CMI.

LFS and CMI data are generated through two separate surveys; hence, the employment figures reported in these two reports are not expected to match. However, the trends in growth should be in the same direction when compared for the same time period. Unfortunately, some serious mismatch can be found as regards the manufacturing employment growth as reported in LFS and those reported by the CMI. The above bar diagram shows the per annum average growth rate of manufacturing employment from CMI and LFS data. It is worth mentioning here that for CMI considers production workers and to make the figures comparable, the CMI figures are taken corresponding to the LFS survey year. It can be seen that during the first (and third) phase of trade liberalisation, both LFS and CMI data shows positive growth with per annum averages being 12.7 and 10.3 (4.3 and 3.1) per cent respectively. In addition, growth rate reported in LFS is higher compared to CMI in both phases. However, in second phase (1991-1996), growth rates according to these two reports differ in their direction; LFS figures show a decline in industrial employment with an average growth of (-) 2.5 per cent whereas CMI have reported a positive growth over the same period with per annum average of 8 per cent.

5.3 Analysis of Public and Private Enterprises

One of the common features of countries pursuing domestic trade reforms has been that they tend to also pursue privatisation and denationalisation policies. Private owned manufacturing industries are given incentives whilst public sector or SOEs in manufacturing sector are downsized and privatised. In all three phases of trade liberalisation, a declining trend in the number of establishments belonging to the public sectors is observed in Bangladesh whilst the number has increased for the private sector.

Figure 5.4

Dynamics of Industrial Employment in Public and Private Sector

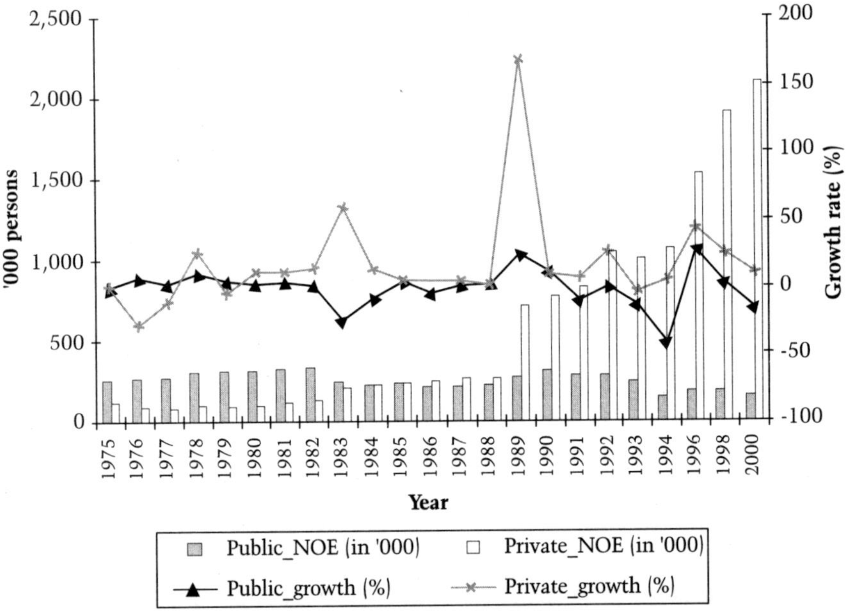

Source: Authors' estimation based on CMI data, various issues.

It is evident from Figure 5.4 that contribution of public sector has declined quite significantly in total manufacturing labour force while private sector shows a rise. In 1975, more than two-thirds of industrial employment was generated by the public sector and up to 1982, this sector posted positive growth with an average share close to 75 per cent. Following this, for most of the years, public sector has experienced a negative growth with the exception of 1989, 1990 and 1996 when more than 20 per cent growth is observed. A sharp decline was observed during phase two when public manufacturing employment figure reached its all time minimum level of 143 thousand. Before 1975, the situation of private manufacturing sector has been quite the opposite because of nationalisation of the large manufacturing industries; in 1975 only 118.5 thousand production workers were employed in private sectors (i.e., 32 per cent of the total industrial employment) and prior to 1979, private sectors experienced a reduction in employment of production workers. Following the introduction of NIP-82 and subsequent industrial policies along with various policy reforms undertaken by the GoB, the growth of private sector employment gathered momentum in early 80s. Highest growth of private manufacturing sector was observed in 1982 (60 per cent) followed by in 1996 (45 per cent).

Figure 5.5

Comparison of Growth Performance of Public and Private Sector

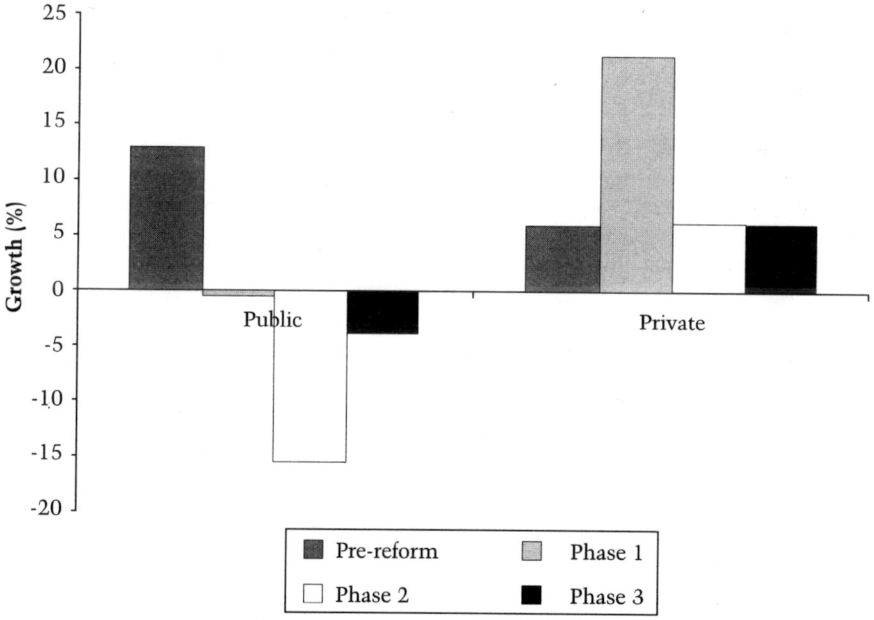

Source: Authors' estimation based on CMI data, various issues.

Figure 5.5 demonstrates phase-wise growth scenario for public sector *vis-à-vis* private sector employment. It is seen that trade liberalisation has adversely affected the public sector manufacturing employment; the second phase has experienced the largest average per annum fall with (-) 15.4 per cent growth. Whilst positive growth has been observed for private sector in

Box 5.3

Closed Down of Public Enterprises

A World Bank study (1994) reveals that by 1974-75, around 305 state owned enterprises (SOEs) comprising industrial, commercial and financial institutions were put under public ownership. However, the size of the public sector enterprises have reduced considerably after the paradigm shift in the government's economic policy towards privatisation. Since the establishment of the Privatization Board in 1993 and thereafter the Privatization Commission in 2000, 74 state owned enterprises were privatised of which 54 were privatised through outright sale and 20 through offloading of shares. The main sectors for privatisation are jute sector, textile sector, steel, engineering and miscellaneous, sugar and food sector and power sector. The following two figures show the number of privatised enterprises and the sectoral distribution of privatised SOEs.

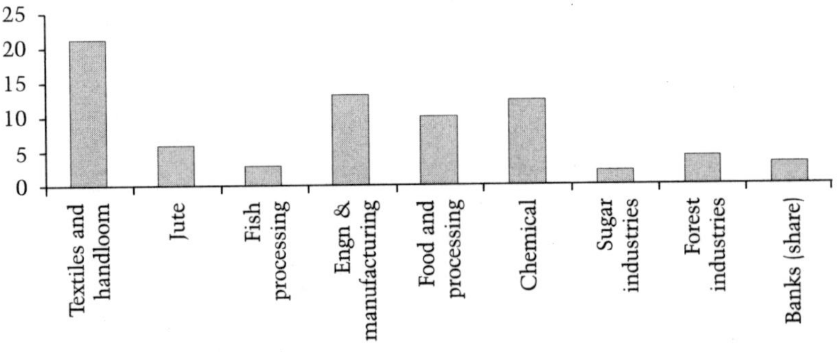

Source: Privatization Commission, Bangladesh.

There has been a trend of privatisation of SOEs from phase 2 resulting in reduction of employment. Recently, Privatization Board has come up with the number of industries to be privatised e.g., for the jute sector the number is 25, for textile, it is 22, 14 for sugar and food sector and 10 for chemical sectors.

Source: Privatization Commission, Bangladesh.

both pre-reform and post-reform era, private sector is found to be the clear winner of the trade reform initiatives with the highest per annum average growth of 21 per cent during the first phase. It appears that the growth in private sector employment stagnated during phases 2 and 3 after a high growth during phase 1. This implies that after an initial burst following the opening up of investment opportunities, growth seems to have tapered off. This would also imply that privatisation and growth of private sector has not really been able to make a significant impact on the growth of manufacturing employment.

5.4 Profile of Manufacturing Sector at 2-Digit Level and Employment

For reporting purpose and to classify the industries, CMI has been using the BSIC codes—the national industry classification system—which has been developed based on the ISIC codes. Four-digit level data is the most disaggregate level at which industry data are reported in CMI that are again grouped to get the three- and two-digit level data.[6] To start with, composition and average share of the manufacturing industries at two-digit level over various phases covering the period 1990-2000 are reported in Table 5.2.[7] To examine the most recent situation, the shares of these industry groups in 2000 have also been presented. Out of the nine two-digit BSIC codes (31-39), textile, wearing apparel and leather industries (BSIC code 32) is the front-runner in terms of various indicators (number of establishment, gross output, VAFC, number of productive workers, capital stock) followed by manufacture of food beverages and tobacco (BSIC code 31). In 2000, these two sectors jointly have constituted 73, 72, 65, 85, 59 and 57 per cent of total establishment, gross output, value added at factor cost, production workers, non-production workers and capital stock respectively, confirming the dominance of these two sectors in the overall manufacturing sector of Bangladesh.

---

6. There are a total of 179 four-digit, 38 three-digit and 9 two-digit BSIC codes.
7. For disaggregate level analysis of CMI data, the time period is chosen as 1990-2000 throughout this study, unless otherwise stated. This period covers two phases of trade liberalisation namely, second and third phase.

However, in case of manufacture of food beverages and tobacco, decrease in shares between second phase and third phase is observed for all reported indicators except capital stock. It is important to note that employment contribution of this sector have decreased five percentage points for production workers and six percentage points for non-production workers between these two phases.

On the other hand, for textile, wearing apparel and leather industries, all indicators show increasing contribution in the third phase compared to the second except non-production workers and the number of establishments. This sector is, by and large, the most significant industrial employment contributor with an average share of more than 80 per cent production workers in the third phase as opposed to 74 per cent in the second phase. However, in the third phase the share of this sector has declined four percentage points with respect to non-production workers.

Careful examination of Table 5.2 reveals that in terms of employment generation, among the remaining eight two-digit level industries, the relative importance of only non-metallic industries (BSIC code 36) has increased for both production and non-production workers. In short, while transitioning to third phase of trade liberalisation from the second, two broad industry sectors (textile, wearing apparel and leather and non metallic industries) have emerged as winners while the relative importance of other sectors have decreased (in terms of employment creation).

One major concern as regards the structure of the manufacturing sector of Bangladesh has been its historically high concentration in few sectors which is also revealed from the above analysis. The disturbing factor is that the manufacturing sector is getting more narrowly based than what has been the situation before trade liberalisation.

5.5 Employment Scenario for Labour-intensive and Export-oriented Industries

The Heckscher-Ohlin (H-O) trade theory postulates that for any country, increased trade raises overall welfare (measured as increased consumption possibilities) by intensifying specialisation in the production of those goods and services, which use relatively abundant factors more

Table 5.2

Sectoral Profile of Manufacturing Industries in Bangladesh (At 2-Digit Level)

BSIC Code (Industry Description)	No. of Establishment				Gross Output				Value Added at Factor Cost			
	Phase 2 (from 1991 to 1996)	Phase 3 (from 1997 to 2000)	2000	After 90	Phase 2	Phase 3	2000	After 90	Phase 2	Phase 3	2000	After 90
31 (Food beverages and tobacco)	23.9	23.1	24.1	23.6	27.3	22.8	21.8	24.7	19.6	15.1	12.8	17.0
32 (Apparel and leather industries)	51.5	50.0	48.9	50.9	42.8	49.0	50.3	46.4	46.1	49.0	52.2	47.8
33 (Wood & wood products, including furniture)	5.2	4.8	3.4	5.0	0.8	0.8	0.8	0.8	1.1	1.1	1.3	1.1
34 (Paper & paper products)	3.2	3.4	3.5	3.3	3.3	3.4	3.7	3.4	3.6	3.8	3.7	3.7
35 (Chemicals & chemical petroleum & rubber)	3.6	3.8	3.9	3.7	11.2	8.7	7.9	9.7	13.5	11.0	9.7	12.0
36 (Non-metallic mineral products)	4.6	7.5	8.6	5.9	2.2	3.0	3.5	2.6	3.1	3.5	4.5	3.4
37 (Basic metal industries)	0.8	0.8	0.8	0.8	5.1	3.7	3.1	4.3	3.1	3.7	2.9	
38 (Fabricated metal production machinery & equipments)	6.4	5.8	6.1	6.2	7.1	8.2	8.6					
39 (Others)	0.9	0.8	0.7	0.9	0.3	0.5	0.3	0.4	0.3	0.4	0.3	0.4
Total	100.0	100.0	100.0	100.0	100.0	100.0	100.0	100.0	100.0	100.0	100.0	100.0
Total in number	26559	27744	24747	27067	1004944	1777090	2005565	1335863	385093	632141	607834	490971

contd...

...contd...

BSIC code (Industry Description)	*Production Worker*				*Non-Production Worker*				*Capital Stock*			
	Phase 2	*Phase 3*	*2000*	*Atfer 90*	*Phase 2*	*Phase 3*	*2000*	*After 90*	*Phase*	*Phase*	*2000*	*After 90*
31 (Food beverages and tobacco)	11.0	6.2	5.7	8.3	35.2	29.0	29.6	31.8	11.6	12.4	14.3	12.0
32 (Apparel and leather industries)	73.9	80.6	79.7	77.8	32.4	28.9	29.8	30.5	30.9	45.1	43.0	38.5
33 (Wood & wood products, including furniture)	1.3	0.8	0.8	1.0	1.4	1.3	1.1	1.4	0.4	0.4	0.3	0.4
34 (Paper & paper products)	2.1	1.9	2.3	2.0	3.2	3.2	4.2	3.2	2.5	1.9	2.0	2.2
35 (Chemicals & chemical petroleum & rubber)	3.2	2.0	1.8	2.5	7.5	6.0	6.8	6.6	36.5	24.0	25.4	29.8
36 (Non-metallic mineral products)	2.5	3.9	5.9	3.3	13.5	26.1	22.5	20.4	2.1	4.0	4.4	3.1
37 (Basic metal industries) 4.1	3.4	1.1	0.7	0.6	0.8	1.6	0.9	0.8	1.2	5.8	2.7	2.2
38 (Fabricated metal production machinery & equipments)	4.6	3.7	3.1	4.1	4.9	4.3	4.8	4.6	10.0	9.0	7.9	9.5
39 (Others)	0.2	0.1	0.1	0.2	0.3	0.3	0.3	0.3	0.1	0.5	0.3	0.3
Total	100.0	100.0	100.0	100.0	100.0	100.0	100.0	100.0	100.0	100.0	100.0	100.0
Total in number	1004944	1777090	2005565	1335863	385093	632141	607834	490971	129616289	200182453	207410724	159858931

Source: Authors' estimation based on CMI data, various issues.

intensively. In short, relatively labour-rich countries export labour-intensive commodities and import the relatively capital-intensive commodities. Hence being a labour surplus country, in case of Bangladesh it would be expected that, due to liberalising trade regime, labour-intensive industries would expand and skill- and/or capital-intensive industries would shrink.

On a different note, tariff liberalisation alters relative incentives to produce for domestic and export markets. Usual hypothesis states that as a consequence of trade liberalisation, import competing sectors will contract, while export-oriented sectors will expand. According to traditional trade theory, the changing pattern of trade will also reflect the economy's relative factor endowments. Different technological capacities and relative availability of resources force each country to identify a set of identifiable exporting and import-competing sectors.

In this section an effort has been made to investigate the employment situation in labour-intensive and export-oriented manufacturing industries. To identify labour-intensive and export-oriented sectors at 4-digit level BSIC code, two indicators namely, capital-labour ratio and export as a percentage of total sales have been examined.

For classifying industries as labour-intensive, in lieu of using the capital-labour ratio of a single year, the yearly average capital-labour ratio for seven years CMI data (from 1991 to 2000) has been considered as the criterion variable. There is no strict rule to define labour-intensive sectors. However, after carefully examining the average capital-labour ratio, an arbitrary threshold (50,000) was selected with an intention to cover at least the lowest one-third industries when these are arranged in ascending order based on capital-labour ratio. In other words, a sector is identified as labour intensive if this average capital-labour ratio is below 50,000 taka/per production worker. A total of 64 sectors (36 per cent of total 179 4-digit industry sectors) satisfy this criterion. An additional sector, ship breaking dismantling (BSIC code 3857), has also been included due to the known characteristic of relatively labour dependence which raised the number of total labour-intensive sectors to 65; the remaining 114 sectors has been identified as non-labour intensive sectors. The full list of labour-intensive sectors with the phase-wise changes in profile is presented in Annex Table A-5.3.

On the other hand, to select export-oriented industries in lieu of using average yearly sectoral exports as a percentage of total sales over 1991-2000, the latest series (i.e., of 2000) was considered to seize the most recent export orientation of the sectors. A sector is identified as export-oriented if export exceeds 20 per cent of total sales. A total of 21 sectors (12 per cent of total 179 4-digit industry sectors) satisfy this criterion. This is consistent with the national export basket and in line with the high concentration of Bangladesh's exports on few products. It is to be noted that five export oriented sectors are also identified as labour-intensive sectors; these are readymade garments (3231), knitting mills (3223), manufacturing of hats and caps (3232), spooling & thread ball (3226) and made up textile goods (3222). The list of export-oriented sectors and phase-wise changes in their respective profiles are presented in Annex Table A-5.4. The remaining 158 sectors are termed as non-export oriented sectors in the subsequent analysis.

Figure 5.6

Dynamics of Production Worker for Major Industry Groups (1991-2000)

Source: Authors' estimation based on CMI data, various issues.

Figure 5.6 displays the dynamics of production workers over the period 1991-2000 for pooled, labour-intensive and export-oriented industries. Similar type of movement in trends for these three groups reveal that for all groups, the employment figures grow at a steady pace up to 1994; a jump is seen in 1996 which is followed by an upward trend till 2000. The employment generated in export-oriented sectors was higher compared to labour-intensive sectors during the second phase of trade liberalisation, more precisely till 1996; for most years during the third phase, employment in labour-intensive sectors was marginally higher than export-oriented sectors.

Figure 5.7

Dynamics of Production Worker for Major Industry Groups (1980-2000)

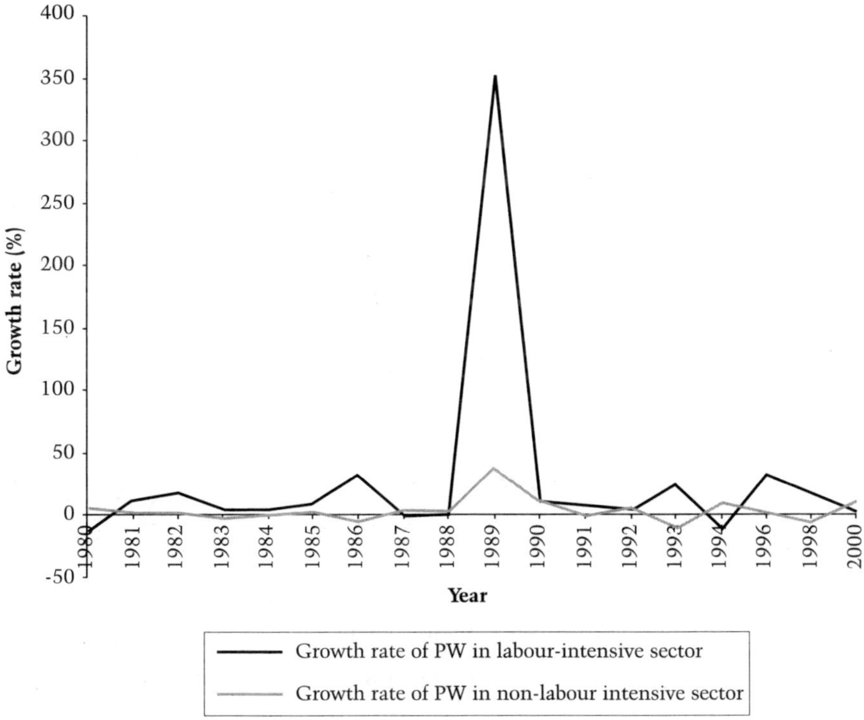

Source: Authors' estimation based on CMI data, various issues.

Both labour and non-labour intensive sectors reveal, in general, a positive growth (Figure 5.7). However, employment market for labour intensive sectors shows more volatility compared to non-labour intensive sectors. An abrupt jump in 1989 is observed for both of these sectors particularly supporting the earlier observation as regards the presence of structural break in the CMI data for 1989. However, for labour-intensive sectors (with more than 350 per cent growth), the impact of this structural break is more visible due to the fact that newly incorporated handloom sector falls in the labour-intensive group. It is to be noted that after a moderate growth during the rapid trade liberalisation regime, including a negative growth of (-) 10.7 per cent in 1994, labour-intensive sectors experienced quite significant increase in employment as a result of which labour force size stood at 1.38 million in 2000 rising from the 0.57 million in 1994 (as shown in Figure 5.6).

Figure 5.8

Growth Rate of Productive Workers in Export and Non-Export Sectors

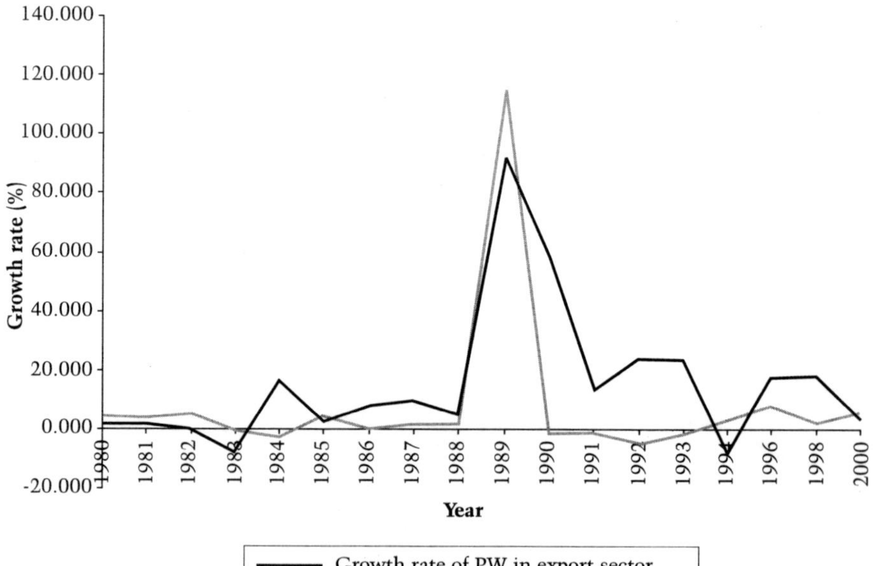

Source: Authors' estimation based on CMI data, various issues.

In case of export-oriented and non-export oriented sectors, number of productive workers have increased more in the former compared to the latter. Figure 5.8 shows growth rates of productive workers in these two sectors. Again, in general terms, employment has experienced positive growth over the period 1980 to 2000. As was seen in earlier cases, unusual growth was observed in 1989 and non-export oriented industries experienced larger jump compared to export-oriented industries.

Similar to labour-intensive sectors, export-oriented sectors experienced highest growth during the third phase. The labour force employed in the export-oriented sector was about half a million at the end of the second phase which phenomenally increased to 1.37 million in 2000.

5.6 Extent of Tariff Liberalisation for Pooled, Labour-intensive and Export-oriented Industries

Aggregate analysis of trade liberalising measures, particularly based on tariff and trade (both export and import) reforms was presented in Chapter 3 to understand the extent, phases and pace of tariff reform. Effective rate of protection in relation to industrial quantum growth performance, for selected export-oriented and import competing/substituting sectors, was also analysed.[8] However, analysis of the tariff protection at disaggregate industry level for Bangladesh is almost non-existent; hence, one of the key value additions of this study to the existing body of literature is to undertake an exercise to bridge this gap. It is rather difficult to map and match I/O sectors and BSIC code. On the other hand, tariff lines are reported following HS nomenclature and a concordance table is also available for ISIC-HS codes in this respect.[9] Again, some discrepancies exist between BSIC and ISIC codes. Therefore, to arrive at sectoral tariff data, correspondence between BSIC and ISIC codes at 4-digit level has to be established first. Consequently, sectoral tariff data series (at 4-digit BSIC)

---

8. Bangladesh Tariff Commission (BTC) released ERP data for various (40) industries and economic sectors, as specified in the I/O table. However, all 179 4-digit BSIC industry sectors are not covered by the I/O table.

9. The concordance between ISIC 4-digit (revision two) and HS 6-digit codes has been extracted from John Haveman's webpage (*http://www.macalester.edu/research/economics/page/haveman/Trade.Resources/tradeconcordances.html#FromISIC*). The original source is the CDROM "A dataset on trade and production 1976-99" that comes along with the book edited by B.Hoekman, A. Mattoo and P. English (2002), *Development, Trade and the WTO: A Handbook*, The World Bank.

were estimated from 1991 to 2000 by using BSIC-ISIC-HS data correspondence. The data source for tariff is National Custom Schedule.

Prior to presenting the analysis of sectoral tariff, the distinction between two measures of tariff used in this study to measure trade restrictions needs to be made. One of these two is customs duty and the other is total tariff incidence (or simply tariff incidence). In Bangladesh, total tariff is charged on importers applying tax on tax system. In other words, the ad valorem custom duty is charged on the declared/assessed import value and the remaining quasi- and para-tariffs (e.g. supplementary duty, VAT etc.) are calculated on add-on basis, not on the initial import value. Thus, the total tariff incidence is calculated by the following formulae:

$$TI=\frac{x+y}{100}+\left\{\left(1+\frac{x+y}{100}\right)*\frac{z}{100}\right\}+\left\{\left(1+\frac{x+y}{100}\right)+\left(1+\frac{x+y}{100}\right)*\frac{z}{100}\right\}*\frac{a}{100}+\frac{b}{100}+\frac{c}{100}$$

Where, x = Custom Duty (CD)

y = Regularity Duty (RD)

z = Supplementary Duty (SD)

a = Value Added Tax (VAT)

b = Advance Income Tax (AIT)

c = Infrastructure Development Search Charge (IDSC)

Whereas custom duty only measures tariff restrictions (MFN applied tariff), tariff incidence measures total level of protection incorporating para-tariff (i.e., supplementary duty, IDSC etc.) and VAT with the MFN applied tariff.

Figure 5.9 (a) and (b) shows the yearly averages of custom duty and tariff incidence (by line graphs) respectively for pooled (all industries), labour-intensive and export-oriented sectors along with their standard errors (by bar graph). It is evident that protection for all three major industry categories, both in terms of custom duty and tariff incidence, increased in 1992 and in subsequent years, the level of protection has posted a monotonic decrease; however, liberalisation took place at a quick pace till 1996 following which it steadied up to a large extent. Export oriented sectors enjoyed more protection compared to pooled and labour intensive sectors in all cases, except 1996 and 1998 when tariff incidence

Figure 5.9

Major Industry Group-wise Extent of Tariff Liberalisation: Average and Dispersion (1991-2000)

(a) Based on Average Custom Duty

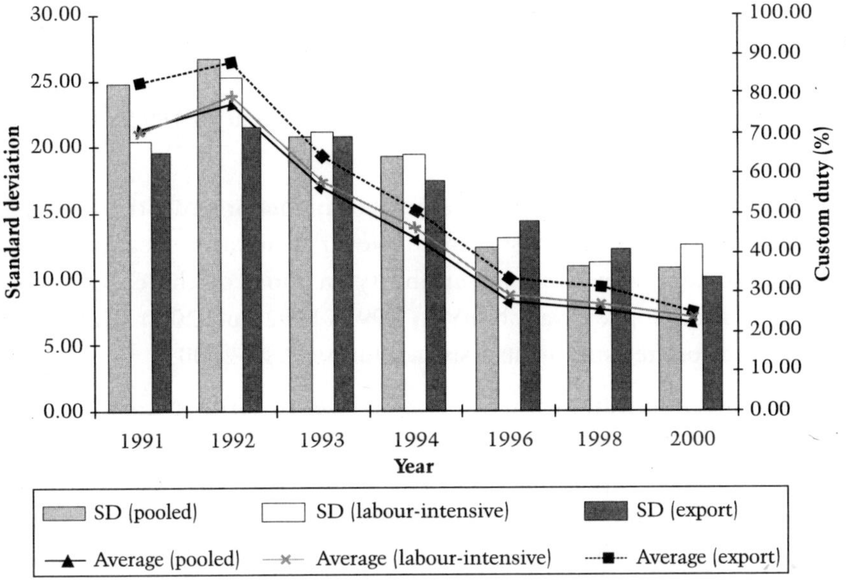

(b) Based on Tariff Incidence

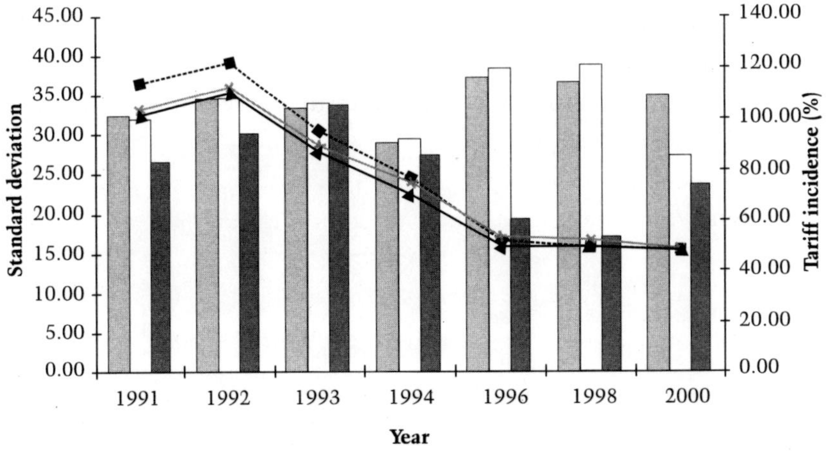

Source: Authors' estimation based on National Custom Schedule and CMI data, various issues using BSIC-ISIC-HS data correspondence.

was marginally higher for labour-intensive than export-oriented sectors. For pooled industries, the custom duty (tariff incidence) was 71.00 (101.60) per cent in 1991 which dropped to 27.65 (50.23) per cent in 1996 and stood at 22.28 (47.97) per cent in 2000. In case of export-oriented industries, the highest average custom duty (tariff incidence) was 88.22 (122.26) per cent in 1992 which gradually declined to 33.56 (51.78) per cent in 1996 and reached to 22.28 (47.97) per cent in 2000. Finally, custom duty (tariff incidence) for labour-intensive industries was 70.43 (103.48) per cent in 1991 which increased to reach 79.67 (112.90) per cent in 1992 which came down to 23.80 (48.52) in 2000.

While examining the dispersion within major industry group, no decisive trend could be observed; however, generally speaking export oriented sectors show lowest variability in most of the cases. Pooled industries show highest variability in 1991, 1992 and 2000 while labour intensive sectors register highest spread during 1993-2000.

Another implication for the lowest level of tariff in case of pooled industry is that non-labour intensive and non-export oriented sectors actually had lower protection compared to their respective counterparts. It may be recalled that growth of production workers in labour-intensive sectors was higher than that of non-labour intensive sectors, while for export-oriented sectors the growth rates tend to be higher, with exceptions seen in particular years. When this is compared with growth analysis of production workers for labour-intensive and export-oriented sectors presented in the previous section, the indication is that protected sectors (labour-intensive and export-oriented sectors, in this case) have relatively higher employment growth in manufacturing sector of Bangladesh compared to the other sectors.

Box 5.4

Analysis of Labour Productivity and Real Wage based on CMI Data: An Anomaly

The figure below shows the movements in labour productivity and real wage indices for overall manufacturing sector based on CMI data. One can see that productivity has increased till 1994 (with a high jump of 31 index points in between 1994 and 1993) and after 1994, a downward trend is observed till 2000. In fact, the productivity index has reached 136 in 1994 and has crawled back to 103 in 2000. This finding is an apparent contradiction with what the LFS data revealed in Chapter 4. The industrial performance judged by QIP index (base year 1988-89=100) also reveals that the index climbed to 214.31 in FY 2000 from 153.83 in 1994 i.e., an upward movement of 60.42 points for overall manufacturing industries during the corresponding period during a period when CMI data shows falling labour productivity (in taka term). In case of real wage index the movement is quite erratic; a fall was observed in 1994 and 1998 with an increase in 1996 and 2000. In contrast, both nominal and real wage indices (provided by Price Section, BBS and reported in SYB) show an increasing trend throughout the corresponding period. The correlation between productivity and wage index is turned out to be insignificant.

Movements in Labour Productivity and Real Wage for Overall Manufacturing Sector (CMI Data: 1991-2000)

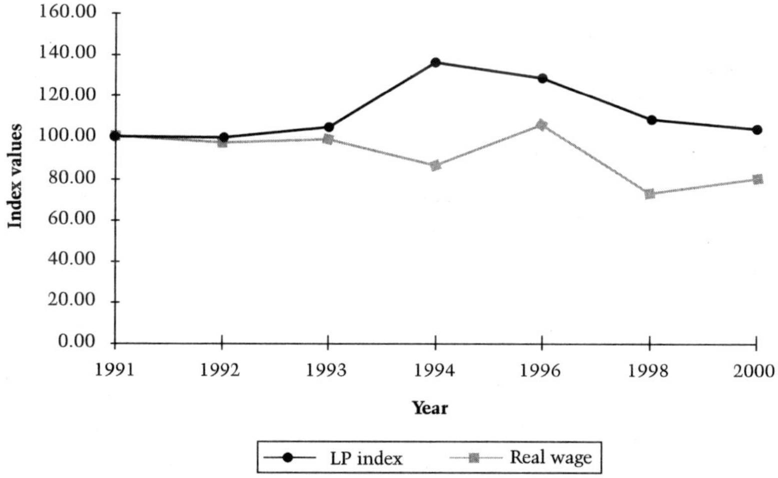

It is widely acceptable and empirically proven that data at aggregate level is more consistent compared to disaggregate data. Due to the suspicious character of the result found in case of overall manufacturing sector, detailed analysis of labour productivity and wage indices and their co-movements at 2-digit and 4-digit level has not been presented here. A correlation analysis between productivity and wage also show insignificant correlation for all two digit industrial sectors except BSIC code 36 (manufacture of non-metallic mineral products) in which case the correlation is 0.77. However, productivity and real wage indices at 2-digit level for all nine sectors have been estimated and presented in the Annex Tables A-5.5. and A-5.6.

Annex A-5.1

Notes on Terminology used in CMI

1. Production workers (PW)

Those who are engaged directly in the production process and include those engaged in manufacturing, assembling, packing, repairing etc. Working supervisors and persons engaged for repair and maintenance are also included. Temporary daily basis hired workers and unpaid family workers are however excluded.

2. All employees (NOE)

Comprise both salaried employees and production workers. Salaried employees include professional and technical workers, administrative and managerial workers, clerical workers, sales and services.

3. Non-production workers (NPW)

It is defined as the difference between all employees and production workers.

4. Wages and salaries

Include all cash payments (monthly salary, house rent, medical allowance and other allowance) made during the year in cash as compensation for work done but does not include other cash and non cash benefits.

5. Value added at factor cost

Census value-added (value of gross output less industrial cost), less non-industrial cost and indirect taxes.

Annex Table A-5.1

Effective Rate of Protection for Export and Import Substituting Industries (%)

Sectors	*1992-93*	*1993-94*	*1994-95*	*1995-96*	*1996-97*	*1997-98*	*1998-99*	*1999-2000*	*2000-01*	*2001-02*	*2002-03*
Sugar	96.30	42.30	52.30	51.10	51.40	40.00	38.50	31.10			
Old ERP (growth)		(-56.07)	(23.64)	(-2.29)	(0.59)	(-22.18)	(-3.75)	(-19.22)			
New ERP (growth)											
Yarn	68.00	57.40	60.90	51.70	35.00	34.20	33.70	30.50			
Old ERP (growth)		(-15.59)	(6.10)	(-15.11)	(-32.30)	(-2.29)	(-1.46)	(-9.50)			
New ERP (growth)		100.90	104.00	96.80	70.00	73.10	81.90	33.40	31.80	32.00	34.10
			(3.07)	(-6.92)	(-27.69)	(4.43)	(12.04)	(-59.22)	(-4.79)	(0.63)	(6.56)
Edible oil	74.80	46.50	39.60	55.60	3.70	41.40	35.30	35.00			
Old ERP (growth)		(-37.83)	(-14.84)	(40.40)	(-93.35)	(1018.92)	(-14.73)	(-0.85)			
New ERP (growth)		99.70	78.10	87.60	90.80	87.20	85.80	52.60	52.80	52.60	40.50
			(-21.66)	(12.16)	(3.65)	(-3.96)	(-1.61)	(-38.69)	(0.38)	(-0.38)	(-23.00)
Paper	68.30	74.10	48.80	25.40	22.70	12.70	11.30	15.50			
Old ERP (growth)		(8.49)	(-34.14)	(-47.95)	(-10.63)	(-44.05)	(-11.02)	(37.17)			
New ERP (growth)		81.20	52.80	26.90	17.70	22.80	22.70	20.70	23.10	21.40	18.10
			(-34.98)	(-49.05)	(-34.20)	(28.81)	(-0.44)	(-8.81)	(11.59)	(-7.36)	(-15.42)
Pharmaceutical	1.50	-2.20	-2.50	-2.60	-1.40	0.70	0.60	-1.70			
Old ERP (growth)		(-246.67)	(13.64)	(4.00)	(-46.15)	(-150.00)	(-14.29)	(-383.33)			
New ERP (growth)		30.30	25.00	23.40	23.40	26.60	23.30	8.00	5.20	5.50	6.70
			(-17.49)	(-6.40)	(0.00)	(13.68)	(-12.41)	(-65.67)	(-35.00)	(5.77)	(21.82)
Cement	56.00	30.60	21.40	18.50	19.10	19.00	20.30	21.20			
Old ERP (growth)		(-45.36)	(-30.07)	(-13.55)	(3.24)	(-0.52)	(6.84)	(4.43)			
New ERP (growth)		20.30	24.30	24.30	24.50	27.00	34.70	43.60	40.10	39.10	50.00
			(19.70)	(0.00)	(0.82)	(10.20)	(28.52)	(25.65)	(-8.03)	(-2.49)	(27.88)
Tobacco cigarette	133.60	69.90	89.70	85.00	86.70	81.90	74.70	68.50			
Old ERP (growth)		(-47.68)	(28.33)	(-5.24)	(2.00)	(-5.54)	(-8.79)	(-8.30)			
New ERP (growth)		-10289.70	-10437.60	-10554.20	-10636.70	-10126.00	-9876.10	3102.50	3106.40	3099.60	4223.60
			(1.44)	(1.12)	(0.78)	(-4.80)	(-2.47)	(-131.41)	(0.13)	(-0.22)	(36.26)
Cloth: mill	189.70	147.50	131.60	98.00	110.20	86.20	78.20	72.70	154.30	62.40	83.50
Old ERP (growth)		(-22.25)	(-10.78)	(-25.53)	(12.45)	(-21.78)	(-9.28)	(-7.03)	(112.24)	(-59.56)	(33.81)
New ERP (growth)		-10289.70	-10437.60	-10544.20	-10636.70	-10126.00	-9876.10	3102.50	3106.40	3099.60	4223.60
			(1.44)	(1.12)	(0.78)	(-4.80)	(-2.47)	(-131.41)	(0.13)	(-0.22)	(36.26)

contd...

...contd...

Sectors	*1992-93*	*1993-94*	*1994-95*	*1995-96*	*1996-97*	*1997-98*	*1998-99*	*1999-2000*	*2000-01*	*2001-02*	*2002-03*
Chemical fertiliser	-5.60	-2.20	-5.00	-3.60	-3.00	0.40	0.50	0.60			
Old ERP (growth)		(-60.71)	(127.27)	(-28.00)	(-16.67)	(-113.33)	(25.00)	(20.00)			
New ERP (growth)		-17.70	-11.80	-8.50	-6.50	-3.40	-2.10	5.80	3.30	3.30	3.80
			(-33.33)	(-27.97)	(-23.53)	(-47.69)	(-38.24)	(-376.19)	(-43.10)	(0.00)	(15.15)
Iron and steel	40.90	27.20	27.40	25.10	24.60	25.00	25.10	19.50			
Old ERP (growth)		(-33.50)	(0.74)	(-8.39)	(-1.99)	(1.63)	(0.40)	(-22.31)			
New ERP (growth)		33.30	27.40	22.90	25.90	27.30	26.70	19.50	23.30	23.30	22.80
			(-17.72)	(-16.42)	(13.10)	(5.41)	(-2.20)	(-26.97)	(19.49)	(0.00)	(-2.15)
Metal products	52.70	43.30	25.10	25.80	27.00	18.20	17.30	15.40			
Old ERP (growth)		(-17.84)	(-42.03)	(2.79)	(4.65)	(-32.59)	(-4.95)	(-10.98)			
New ERP (growth)		41.60	22.30	24.50	22.00	23.20	22.50	23.30	25.80	26.00	21.00
			(-46.39)	(9.87)	(-10.20)	(5.45)	(-3.02)	(3.56)	(10.73)	(0.78)	(-19.23)
Machinery	47.50	28.90	15.10	12.60	12.30	9.30	9.60	5.20			
Old ERP (growth)		(-39.16)	(-47.75)	(-16.56)	(-2.38)	(-24.39)	(3.23)	(-45.83)			
New ERP (growth)		22.00	0.90	1.70	3.20	4.20	4.80	7.70	5.90	6.20	8.20
			(-95.91)	(88.89)	(88.24)	(31.25)	(14.29)	(60.42)	(-23.38)	(5.08)	(32.26)
Transport equipment	69.90	49.10	41.90	38.00	22.80	21.80	19.80	17.90			
Old ERP (growth)		(-29.76)	(-14.66)	(-9.31)	(-40.00)	(-4.39)	(-9.17)	(-9.60)			
New ERP (growth)		43.10	38.30	37.60	35.30	35.20	34.50	42.30	32.00	32.30	39.60
			(-11.14)	(-1.83)	(-6.12)	(-0.28)	(-1.99)	(22.61)	(-24.35)	(0.94)	(22.60)
Jute		64.40	67.20	30.60	31.80	32.00	32.40	32.50	26.80		
Old ERP (growth)			(4.35)	(-54.46)	(3.92)	(0.63)	(1.25)	(0.31)	(-17.54)		
New ERP (growth)			70.80	33.90	35.30	35.40	38.00	38.10	60.10	34.80	34.70
				(-52.12)	(4.13)	(0.28)	(7.34)	(0.26)	(57.74)	(-42.10)	(-0.29)
Tea	82.70	85.10	66.00	48.70	48.80	46.40	43.40	41.00			
Old ERP (growth)		(2.90)	(-22.44)	(-26.21)	(0.21)	(-4.92)	(-6.47)	(-5.53)			
New ERP (growth)		123.70	100.00	74.20	73.50	72.00	67.40	63.60	64.10	64.10	53.10
			(-19.16)	(-25.80)	(-0.94)	(-2.04)	(-6.39)	(-5.64)	(0.79)	(0.00)	(-17.16)
Fish and fish products	78.40	45.30	45.10	28.40	28.50	27.90	28.00	23.20			
Old ERP (growth)		(-42.22)	(-0.44)	(-37.03)	(0.35)	(-2.11)	(0.36)	(-17.14)			
New ERP (growth)		38.50	46.60	47.20	47.30	50.00	50.50	38.40	37.70	53.20	43.90
			(21.04)	(1.29)	(0.21)	(5.71)	(1.00)	(-23.96)	(-1.82)	(41.11)	(-17.48)

contd...

...contd...

Sectors	*1992-93*	*1993-94*	*1994-95*	*1995-96*	*1996-97*	*1997-98*	*1998-99*	*1999-2000*	*2000-01*	*2001-02*	*2002-03*
Readymade garments	237.20	130.00	84.10	53.70	57.40	65.40	60.50	58.90			
Old ERP (growth)		(-45.19)	(-35.31)	(-36.15)	(6.89)	(13.94)	(-7.49)	(-2.64)			
New ERP (growth)		135.50	77.30	22.50	23.20	22.30	18.30	50.30	42.80	42.70	40.40
			(-42.95)	(-70.89)	(3.11)	(-3.88)	(-17.94)	(174.86)	(-14.91)	(-0.23)	(-5.39)
Jute textile	98.20	93.50	81.00	53.70	56.00	48.40	44.10	43.50			
Old ERP (growth)		(-4.79)	(-13.37)	(-33.70)	(4.28)	(-13.57)	(-8.88)	(-1.36)			
New ERP (growth)		89.80	80.50	55.40	54.50	52.30	47.40	64.40	192.70	248.10	80.20
			(-10.36)	(-31.18)	(-1.62)	(-4.04)	(-9.37)	(35.86)	(199.22)	(28.75)	(-67.67)
Leather and leather products	98.60	87.30	42.30	20.70	15.00	8.80	5.90	6.50			
Old ERP (growth)		(-11.46)	(-51.55)	(-51.06)	(-27.54)	(-41.33)	(-32.95)	(10.17)			
New ERP (growth)		74.20	63.60	71.10	74.40	75.50	71.80	133.30	125.50	148.30	103.40
			(-14.29)	(11.79)	(4.64)	(1.48)	(-4.90)	(85.65)	(-5.85)	(18.17)	(-30.28)
Average											
Old ERP (growth)	75.71	56.65	40.57	33.02	32.42	28.55	26.77	24.47			
		(-25.18)	(-28.38)	(-18.61)	(-1.82)	(-11.94)	(-6.23)	(-8.59)			

Note: Figures in parenthesis indicate growth rate over previous year (per cent).

Source: Author's estimation based on data from Bangladesh Tariff Commission.

Annex Table A-5.2

Industrial Production Growth in Terms of Quantum Index (Base Year: 1988-89=100)

Sectors	*1990-91*	*1991-92*	*1992-93*	*1993-94*	*1994-95*	*1995-96*	*1996-97*	*1997-98*	*1998-99*	*1999-2000*	*2000-01*	*2001-02*	*2002-03*
Sugar (growth %)	224.32	178.84 (-20.27)	170.89 (-4.45)	201.76 (18.06)	246.22 (22.04)	167.75 (-31.87)	123.33 (-26.48)	151.71 (23.01)	139.08 (-8.33)	112.49 (-19.12)	89.16 (-20.74)	186.23 (108.87)	161.69 (-13.18)
Yarn	109.63	116.49 (6.26)	112.11 (-3.76)	103.15 (-7.99)	84.47 (-18.11)	85.16 (.82)	102.22 (20.03)	107.76 (5.42)	111.68 (3.64)	119.3 (6.82)	123.95 (3.90)	133.64 (7.82)	142.33 (6.50)
Edible oil	104.08	104.51 (0.41)	96.69 (-7.48)	62.79 (-35.06)	56.82 (-9.51)	58.56 (3.06)	49.79 (-14.98)	51.54 (3.51)	49.55 (-3.86)	59.3 (19.68)	59.66 (0.61)	60.13 (.79)	61.85 (2.86)
Paper	105.15	103.08 (-1.97)	104.74 (1.61)	105.41 (.0.64)	99.16 (-5.93)	96.11 (-3.08)	78.78 (-18.03)	53.58 (-31.99)	69.85 (30.37)	64.15 (-8.16)	59.7 (-6.94)	47.11 (-21.09)	35.26 (-25.15)
Pharmaceutical	128.40	158.05 (23.09)	202.84 (28.34)	217.91 (7.43)	264.79 (21.51)	282.56 (6.71)	314.73 (11.390)	309.09 (-1.79)	312.59 (1.13)	301.65 (-3.50)	352.66 (16.91)	418.39 (18.64)	470.21 (12.39)
Cement	79.81	79.2 (-.76)	60.31 (-23.85)	94.04 (55.93)	91.98 (-2.19)	123.87 (34.67)	177.47 (43.27)	157.8 (-11.08)	227.55 (44.20)	399.39 (75.52)	486.18 (21.73)	502.78 (3.41)	510.4 (1.52)
Tobacco cigarette	87.23	88.96 (1.98)	81.74 (-8.12)	89.83 (9.90)	123.36 (37.33)	115.15 (-6.66)	132.03 (14.66)	141.17 (6.92)	138.82 (-1.66)	140.06 (.89)	142.82 (1.97)	144.69 (1.31)	159.7 (10.37)
Cloth: mill	92.81	91.01 (-1.94)	69.74 (-23.37)	48.78 (-30.05)	26.26 (-46.17)	15.8 (-39.83)	16.86 (6.71)	15.85 (-5.99)	17.24 (8.77)	19.18 (11.25)	22.78 (18.77)	24.95 (9.53)	27.88 (11.74)
Chemical fertiliser	95.93	108.57 (13.18)	128.28 (18.15)	148.02 (15.39)	133.01 (-10.14)	140.63 (5.73)	110.89 (-21.15)	127.03 (14.55)	112.56 (-11.39)	199.11 (76.89)	129.71 (-34.86)	109.7 (-15.43)	141.6 (29.08)
Iron and steel	75.03	57.48 (-23.39)	62.03 (7.92)	109.82 (77.04)	182.9 (66.55)	159.18 (-12.97)	172.55 (8.40)	186.42 (8.04)	163.33 (-12.39)	170.06 (4.12)	188.35 (10.76)	197.29 (4.75)	204.81 (3.81)
Metal products	91.34	93.01 (1.83)	90.17 (-3.05)	89.08 (-1.21)	99.99 (12.25)	90.83 (-9.16)	95.76 (5.43)	102.65 (7.20)	107.17 (4.40)	97.96 (-8.59)	102.84 (4.98)	108.63 (5.63)	109.42 (0.73)
Machinery	111.58	114.88 (2.96)	121.00 (5.33)	65.64 (-45.75)	35.43 (-46.02)	36.41 (2.77)	48.33 (32.74)	74.76 (54.69)	50.69 (-32.20)	47.94 (-5.43)	65.56 (36.75)	67.00 (2.20)	73.81 (10.16)
Transport equipment	95.81	96.39 (0.61)	82.81 (-14.09)	62.21 (-24.88)	73.36 (17.92)	67.40 (-8.12)	67.74 (0.50)	75.71 (11.77)	71.43 (-5.65)	71.46 (0.04)	59.92 (-16.15)	64.60 (7.81)	54.56 (-15.54)

contd...

...contd...

Sectors	*1990-91*	*1991-92*	*1992-93*	*1993-94*	*1994-95*	*1995-96*	*1996-97*	*1997-98*	*1998-99*	*1999-2000*	*2000-01*	*2001-02*	*2002-03*
Jute	53.02	52.73 (-0.54)	49.19 (-6.71)	44.53 (-9.49)	53.10 (19.24)	40.74 (-23.26)	48.66 (19.44)	58.24 (19.68)	44.75 (-23.16)	39.21 (-12.40)	45.26 (15.45)	47.33 (4.58)	44.08 (-6.88)
Tea	106.61	109.41 (2.63)	118.47 (8.28)	123.53 (4.27)	112.93 (-8.58)	123.65 (9.49)	127.83 (3.38)	128.60 (0.60)	105.11 (-18.27)	123.52 (17.51)	127.68 (3.37)	134.36 (5.23)	130.40 (-2.95)
Fish and fish products		108.96	125.20 (14.90)	143.64 (14.73)	171.07 (19.10)	169.43 (-0.96)	167.53 (-1.12)	121.25 (-27.62)	129.47 (6.78)	186.14 (43.77)	192.97 (3.67)	201.28 (4.31)	166.73 (-17.17)
Readymade garments	157.91	230.55 (46.00)	286.94 (24.46)	273.56 (-4.66)	280.95 (2.70)	439.98 (56.60)	507.17 (15.27)	644.89 (27.15)	710.61 (10.19)	766.32 (7.84)	811.67 (5.92)	770.28 (-5.10)	804.94 (4.50)
Jute textile	85.14	81.69 (-4.05)	87.52 (7.14)	82.75 (-5.45)	83.46 (0.86)	79.46 (-4.79)	80.13 (0.84)	80.59 (0.57)	72.19 (-10.42)	67.06 (-7.11)	66.78 (-0.42)	69.04 (3.38)	62.80 (-9.04)
Leather and leather products	84.34	92.97 (10.23)	108.72 (16.94)	123.44 (13.54)	126.77 (2.70)	133.85 (5.58)	100.84 (-24.66)	102.25 (1.40)	136.72 (33.71)	154.41 (12.94)	165.85 (7.41)	146.37 (-11.75)	128.64 (-12.11)

Note: Figures in the parentheses indicate growth rate over previous years (in per cent).

Source: Author's estimation based on BBS.

Annex Table A-5.3

Profile of Labour-intensive Industries: Phase-wise Sectoral Distribution (in %) of Selected Key Industrial Variables

BSIC Code	Name of the Industry	Capital-labour Ratio ('000 TK/ worker)	Production Worker				Gross Output				Value-added at Factor Cost				Capital Stock			
			PR	P-1	P-2	P-3	PR	P-1	P-2	P-3	PR	P-1	P-2	P-3	PR	P-1	P-2	P-3
3912	Paper product handicrafts	0.12	0.00	0.01	0.01	0.00	0.00	0.00	0.00	0.00	0.00	0.00	0.00	0.00	0.00	0.00	0.00	0.00
3426	Metal steels printing	2.51	0.00	0.00	0.00	0.00	0.00	0.00	0.00	0.00	0.00	0.00	0.00	0.01	0.00	0.00	0.00	0.00
3271	Embroidery of textile	3.35	0.00	0.10	0.10	0.00	0.00	0.00	0.00	0.00	0.00	0.01	0.02	0.02	0.00	0.01	0.02	0.02
3858	Animal & hand drawn carts	3.94	0.00	0.00	0.00	0.00	0.00	0.00	0.00	0.00	0.00	0.00	0.00	0.00	0.00	0.00	0.00	0.00
3941	Signs & advertise display	5.43	0.00	0.01	0.02	0.01	0.00	0.00	0.01	0.00	0.00	0.01	0.01	0.01	0.00	0.00	0.01	0.00
3911	Wood bamboo handicrafts	5.61	0.00	0.00	0.01	0.00	0.00	0.00	0.00	0.00	0.00	0.00	0.01	0.00	0.00	0.00	0.00	0.00
3915	Metal decorative handicraft	6.04	0.00	0.00	0.00	0.00	0.00	0.00	0.00	0.00	0.00	0.00	0.00	0.00	0.00	0.00	0.00	0.00
3318	Bamboo, cane product NEC	6.56	0.00	0.56	0.70	0.15	0.00	0.06	0.08	0.05	0.00	0.15	0.21	0.14	0.00	0.07	0.14	0.09
3943	Bangles (ex-precious metal	6.72	0.00	0.04	0.02	0.01	0.00	0.03	0.01	0.00	0.00	0.02	0.01	0.00	0.00	0.02	0.01	0.00
3519	Medicinal & pharma. prepn.	6.81	0.00	0.00	0.00	0.00	0.00	0.00	0.00	0.00	0.00	0.00	0.00	0.00	0.00	0.00	0.00	0.00
3425	Book binding & other artc	6.99	0.21	0.08	0.08	0.09	0.23	0.04	0.02	0.01	0.15	0.06	0.06	0.02	0.08	0.03	0.05	0.01
3824	Tin cans & tinware	8.05	1.22	0.41	0.13	0.04	2.48	0.81	0.15	0.04	1.62	0.77	0.21	0.04	3.39	1.03	0.08	0.01
3216	Handloom textiles	8.40	42.88	35.56	27.40	15.16	22.18	21.42	13.93	4.53	20.87	21.81	15.81	6.27	14.84	14.38	16.23	5.57

contd...

...contd...

			PR	P-1	P-2	P-3	PR	P-1	P-2	P-3	PR	P-1	P-2	P-3	PR	P-1	P-2	P-3
3825	Metal trunks	8.93	0.25	0.06	0.02	0.01	0.20	0.05	0.01	0.00	0.20	0.06	0.01	0.00	0.16	0.04	0.01	0.00
3316	Cork and its products	10.11	0.00	0.00	0.00	0.00	0.00	0.00	0.00	0.00	0.00	0.00	0.00	0.00	0.00	0.01	0.00	0.00
3913	Rubber & plastic handrft.	10.33	0.00	0.00	0.00	0.00	0.00	0.00	0.00	0.00	0.00	0.00	0.00	0.00	0.00	0.00	0.00	0.00
3526	Resins plastic materials	10.52	0.00	0.02	0.01	0.01	0.00	0.10	0.00	0.01	0.00	0.12	0.00	0.01	0.00	0.12	0.01	0.00
3142	Cheroots manufacturing	11.09	0.01	0.12	0.02	0.00	0.01	0.00	0.00	0.01	0.01	0.00	0.01	0.01	0.09	0.02	0.01	0.00
3313	Wooden structural product	11.14	0.00	0.03	0.05	0.02	0.00	0.03	0.03	0.01	0.00	0.03	0.04	0.02	0.00	0.02	0.03	0.01
3859	Aircraft,trans equip NEC	12.94	0.00	0.00	0.00	0.00	0.00	0.00	0.00	0.00	0.00	0.00	0.00	0.00	0.00	0.00	0.00	0.00
3424	Printed cards & stationary	13.12	0.00	0.02	0.03	0.03	0.00	0.03	0.05	0.06	0.00	0.02	0.04	0.03	0.00	0.01	0.02	0.03
3232	Mfg. of hats and caps	13.70	0.08	0.00	0.00	0.01	0.12	0.00	0.00	0.01	0.11	0.00	0.00	0.01	0.06	0.00	0.00	0.01
3619	Pottery, china, earthen NEC	15.08	1.47	0.86	0.02	0.01	1.38	1.06	0.01	0.01	1.33	1.24	0.02	0.01	4.07	5.02	0.02	0.01
3231	Readymade garments	15.64	0.06	33.20	51.47	71.28	0.11	34.41	56.91	71.93	0.07	30.33	52.96	67.53	0.03	25.53	45.03	69.65
3836	Computing and accounting machine	17.02	0.00	0.00	0.30	0.16	0.00	0.00	0.90	0.48	0.00	0.00	0.43	0.29	0.00	0.00	1.56	0.78
3537	Candle manufacturing	17.11	0.12	0.02	0.00	0.00	0.08	0.02	0.00	0.00	0.05	0.02	0.00	0.00	0.11	0.02	0.00	0.01
3872	Optical goods	17.51	0.08	0.00	0.03	0.01	0.04	0.00	0.02	0.01	0.05	0.00	0.03	0.01	0.15	0.00	0.05	0.01
3323	Cane & bamboo furniture	17.60	0.00	0.00	0.02	0.01	0.00	0.00	0.01	0.00	0.00	0.00	0.01	0.01	0.00	0.00	0.02	0.00
3429	Photo typesetting	20.96	0.00	0.00	0.00	0.00	0.00	0.00	0.00	0.00	0.00	0.00	0.00	0.00	0.00	0.00	0.00	0.00

contd...

...contd...

			PR	*P-1*	*P-2*	*P-3*	*PR*	*P-1*	*P-2*	*P-3*	*PR*	*P-1*	*P-2*	*P-3*	*PR*	*P-1*	*P-2*	*P-3*
3691	Bricks, tiles & clay prod.	22.41	2.74	3.37	3.67	4.86	2.62	2.30	2.69	2.79	4.40	3.86	6.33	5.68	8.40	3.68	4.08	6.83
3821	Utensils copper/ brass	23.52	0.00	0.06	0.05	0.13	0.00	0.04	0.03	1.28	0.00	0.05	0.04	5.07	0.00	0.04	0.07	0.25
3223	Knitting mills	24.62	3.27	2.93	2.04	2.94	5.58	5.93	3.26	4.95	4.61	3.63	3.35	2.72	4.96	4.75	2.99	3.69
3261	Ginning & baling of cotton	25.29	0.00	0.01	0.01	0.01	0.00	0.01	0.00	0.00	0.00	0.00	0.01	0.00	0.00	0.01	0.02	0.01
3933	Musical instruments	25.68	0.00	0.01	0.01	0.00	0.00	0.00	0.00	0.01	0.00	0.00	0.00	0.01	0.00	0.00	0.01	0.02
3837	Sewing machines	25.94	0.13	0.05	0.25	0.15	0.10	0.22	0.46	0.26	0.07	0.21	0.94	0.37	0.19	0.31	4.24	1.62
3935	Toys, non-power driven	26.75	0.00	0.02	0.02	0.00	0.00	0.01	0.01	0.00	0.00	0.01	0.01	0.00	0.00	0.05	0.03	0.00
3535	Matches manufacturing	26.86	23.54	5.61	1.14	0.37	12.73	5.49	0.81	0.40	16.85	8.74	1.12	0.41	9.33	6.85	1.85	0.86
3321	Wooden furniture	28.04	2.68	1.33	0.48	0.39	2.14	1.60	0.79	1.38	3.06	2.06	1.56	1.83	3.76	2.07	0.68	0.72
3811	Cutlery	29.15	0.38	0.20	0.07	0.04	0.41	0.24	0.08	0.06	0.57	0.24	0.14	0.07	2.81	0.63	0.12	0.06
3242	Fur dressing and dyeing	30.91	0.00	0.01	0.00	0.00	0.00	0.41	0.00	0.00	0.00	0.09	0.00	0.00	0.00	0.00	0.00	0.00
3512	Unani medicines	31.70	0.70	0.22	0.02	0.01	0.36	0.25	0.02	0.01	0.43	0.70	0.03	0.01	0.65	0.40	0.02	0.03
3143	*Bidies*	31.88	0.00	3.11	5.42	0.46	0.00	3.97	6.23	2.32	0.00	6.88	6.77	1.89	0.00	1.32	7.24	1.12
3812	Hand and edge tools	33.04	0.18	0.17	0.07	0.05	0.10	0.21	0.04	0.04	0.09	0.18	0.06	0.04	0.16	0.46	0.11	0.14
3712	Iron and steel foundries	34.66	1.42	0.20	0.07	0.04	2.60	0.64	0.09	0.24	1.29	0.36	0.06	0.20	1.95	0.40	0.22	0.02
3319	Wood & wood cork prods	35.48	0.00	0.03	0.01	0.00	0.00	0.02	0.00	0.00	0.00	0.03	0.01	0.00	0.00	0.02	0.01	0.01
3215	Narrow fabrics	37.34	0.59	0.23	0.19	0.12	0.52	0.21	0.10	0.10	0.50	0.26	0.14	0.12	0.96	0.65	0.43	0.17
3312	Plywood & plywood product	37.58	1.47	0.51	0.14	0.05	2.43	0.97	0.16	0.04	1.98	1.02	0.24	0.08	0.98	1.16	0.47	0.04

contd...

...contd...

			PR	*P-1*	*P-2*	*P-3*	*PR*	*P-1*	*P-2*	*P-3*	*PR*	*P-1*	*P-2*	*P-3*	*PR*	*P-1*	*P-2*	*P-3*
3151	Prepared animal feeds	38.06	0.00	0.02	0.00	0.00	0.00	0.04	0.00	0.00	0.00	0.04	0.00	0.00	0.00	0.59	0.00	0.00
3315	Struct products of bamboo	38.32	0.00	0.01	0.01	0.00	0.00	0.00	0.00	0.00	0.00	0.01	0.01	0.00	0.00	0.01	0.05	0.00
3611	Earthenware	39.27	0.00	0.08	0.06	0.03	0.00	0.05	0.03	0.01	0.00	0.09	0.10	0.03	0.00	0.15	0.09	0.04
3693	Cement products	39.35	0.10	0.07	0.02	0.01	0.14	0.10	0.03	0.02	0.14	0.19	0.05	0.03	0.18	0.93	0.07	0.01
3514	Homeopathic & bio-chemic	41.43	0.52	0.14	0.03	0.01	0.50	0.22	0.02	0.01	0.55	0.36	0.05	0.02	0.43	0.28	0.07	0.03
3819	Utensils-aluminium	41.62	6.97	3.31	1.27	0.59	11.02	4.44	1.77	0.59	8.54	3.36	1.34	0.53	9.34	4.71	3.50	1.11
3871	Mfg. of photographic good	41.87	0.06	0.05	0.04	0.01	0.04	0.05	0.01	0.00	0.05	0.07	0.03	0.00	0.02	0.03	0.10	0.03
3145	*Zarda* and quivam	41.99	0.16	0.07	0.04	0.01	0.56	0.14	0.03	0.01	0.31	0.18	0.03	0.02	0.33	0.20	0.03	0.04
3513	*Ayurvedic* medicines	42.36	1.82	0.60	0.16	0.08	4.13	1.17	0.18	0.19	6.24	1.46	0.23	0.42	2.49	1.65	0.41	0.20
3124	Manufacture of *gur*	44.19	0.00	0.02	0.01	0.01	0.00	0.02	0.01	0.00	0.00	0.03	0.02	0.01	0.00	0.02	0.03	0.01
3226	Spooling & thread ball	44.68	0.98	0.30	0.03	0.03	1.17	0.39	0.02	0.04	0.85	0.35	0.02	0.03	1.87	0.93	0.08	0.12
3931	Jewellery precious metal	45.33	0.00	0.03	0.07	0.02	0.00	0.01	0.03	0.01	0.00	0.03	0.11	0.03	0.00	0.01	0.20	0.05
3222	Made-up textile goods	46.35	0.00	0.39	0.39	0.20	0.00	0.38	0.37	0.67	0.00	0.57	0.56	0.72	0.00	0.41	0.82	0.76
3534	Mfg. of polishes and waxes	48.38	0.02	0.01	0.00	0.00	0.05	0.02	0.00	0.00	0.01	0.02	0.00	0.00	0.08	0.02	0.00	0.01
3122	Bakery products	48.50	3.51	3.78	2.41	1.10	17.82	8.38	4.13	2.37	12.27	7.11	4.15	2.25	16.86	16.19	5.36	4.19
3128	Extraction of edible salt	48.62	0.00	1.21	0.64	0.68	0.00	1.34	1.70	0.97	0.00	1.20	1.73	0.82	0.00	1.34	2.16	1.11
3841	Elec. and machine apparat.	48.91	0.00	0.09	0.04	0.03	0.00	0.31	0.12	0.08	0.00	0.49	0.10	0.09	0.00	0.62	0.12	0.10

contd...

...contd...

			PR	P-1	P-2	P-3	PR	P-1	P-2	P-3	PR	P-1	P-2	P-3	PR	P-1	P-2	P-3
3857	Ship breaking dismantling	58.82	2.37	0.64	0.69	0.54	8.12	2.36	4.62	3.96	12.73	1.46	0.79	2.04	11.25	2.78	1.01	0.39
Total of labour-intensive sectors (in unit**)		Na	100 (34)	100 (112)	100 (489)	100 (1205)	100 (7636)	100 (21351)	100 (87920)	100 (207866)	100 (2177)	100 (5370)	100 (19990)	100 (46545)	100 (1474)	100 (3175)	100 (7958)	100 (20884)
Share of labour-intensive sectors as % of all industry		Na	10.54	24.09	48.41	67.68	5.15	10.73	29.78	39.14	5.96	10.61	28.88	36.73	3.72	4.17	6.17	10.32
All industry in unit**		Na	318	464	1009	1781	148316	199039	295184	531074	36511	50613	69227	126711	39593	76162	128961	202282

Note: **Unit for production workers in 000; others are in million taka.

1. Labour-intensive sectors are identified based on average capital-labour ratio over the years 1991-2000. A sector is identified as labour-intensive if this average capital-labour ratio is below 50,000 taka/per production worker. A total of 64 sectors (36 per cent of total 179 4-digit industry sectors) satisfy this criterion. An additional sector 3857 (Ship breaking dismantling) is also included due to the known characteristic of relatively labour dependence to raise the number of labour-intensive sectors at 65.
2. PR, P-1, P-2 & P-3 means Pre-reform (1978-1980), Phase 1 (1981-1990), Phase 2 (1991-1995) & Phase 3 (1996-2000) respectively.
3. The industries are arranged according to the ascending order of capital-labour ratio.

Source: Author's estimation based on CMI reports, various years.

Annex Table A-5.4

Profiles of Export-oriented Industries: Phase-wise Sectoral Distribution (in %) of Selected Key Industrial Variables

BSIC Code	Name of the Industry	Export % Total Sale	Export P-2	Export P-3	Production Worker PR	Production Worker P-1	Production Worker P-2	Production Worker P-3	Gross Output PR	Gross Output P-1	Gross Output P-2	Gross Output P-3	Value-added at Factor Cost PR	Value-added at Factor Cost P-1	Value-added at Factor Cost P-2	Value-added at Factor Cost P-3
3114	Fish and sea foods	98.98	11.73	8.21	0.21	0.73	0.42	0.36	2.27	7.87	8.16	6.17	0.67	3.36	3.29	3.42
3127	Tea and coffee blending	100.00	0.49	0.13	0.10	0.13	0.03	0.03	1.79	1.32	0.73	0.63	0.57	0.86	0.41	0.13
3152	Bone crushing	38.95	0.00	0.01	0.02	0.02	0.01	0.00	0.03	0.05	0.02	0.02	0.02	0.05	0.02	0.03
3211	Cotton textiles	23.04	0.27	1.85	26.02	18.40	19.84	7.94	25.03	17.68	16.46	8.47	21.28	22.01	20.12	8.03
3213	Jute textiles	89.66	17.56	5.92	69.36	61.91	29.14	12.44	53.02	40.01	15.71	6.67	70.95	53.69	24.68	11.92
3222	Made-up textile goods	86.86	0.10	0.36	0.00	0.15	0.32	0.20	0.00	0.12	0.26	0.53	0.00	0.15	0.39	0.59
3223	Knitting mills	78.27	0.18	3.44	0.52	1.17	1.68	2.91	0.75	1.82	2.34	3.97	0.55	0.94	2.30	2.22
3224	Carpets and rugs	31.21	0.60	0.15	0.01	0.43	0.48	0.34	0.01	0.50	0.90	0.43	0.00	0.67	1.39	0.85
3226	Spooling & thread ball	96.23	0.00	0.04	0.16	0.12	0.03	0.03	0.16	0.12	0.02	0.03	0.10	0.09	0.01	0.02
3231	Readymade garments	99.57	55.33	71.63	0.02	13.26	42.31	70.60	0.03	10.56	40.87	57.65	0.02	7.87	36.31	55.19
3232	Mfg. of hats and caps	100.00	0.00	0.02	0.00	0.00	0.00	0.01	0.00	0.00	0.00	0.01	0.00	0.00	0.00	0.01
3241	Tanning & finishing	99.67	10.20	4.85	0.87	1.21	1.33	0.73	9.85	11.79	8.18	4.30	1.70	3.88	3.79	1.25
3251	Leather footwear	93.10	0.91	2.56	0.27	0.41	2.60	3.44	1.07	2.19	2.47	9.52	0.72	2.63	2.98	14.53
3263	Jute pressing & baling	100.00	1.84	0.01	0.35	0.51	0.43	0.20	3.23	3.26	1.56	0.10	1.34	1.44	1.52	0.26
3412	Paper board manufacturing	27.79	0.26	0.20	0.11	0.20	0.35	0.14	0.18	0.49	0.74	0.58	0.11	0.23	1.05	0.53
3413	Articles of pulp, paper	71.04	0.00	0.27	0.38	0.30	0.11	0.22	0.65	0.64	0.29	0.35	0.51	0.73	0.30	0.37
3419	Manufac. of paper N.E.C	66.07	0.00	0.05	0.00	0.01	0.03	0.03	0.00	0.02	0.05	0.07	0.00	0.02	0.05	0.05
3612	China and ceramic	42.52	0.51	0.24	0.45	0.58	0.84	0.35	0.37	0.81	1.20	0.45	0.39	0.83	1.37	0.49
3822	Utensils steel	77.43	0.00	0.01	1.11	0.40	0.03	0.01	1.48	0.66	0.01	0.01	1.01	0.44	0.02	0.02
3939	Button studs hooks fasten	43.32	0.00	0.01	0.00	0.01	0.01	0.02	0.00	0.01	0.01	0.02	0.00	0.01	0.01	0.04
3944	Othr Mfg (brooms/brushes)	100.00	0.00	0.03	0.03	0.02	0.01	0.01	0.05	0.09	0.01	0.02	0.05	0.08	0.01	0.05
Total of export-oriented sectors (in unit**)			100.00 (72287)	100.00 (157154)	100.00 (211)	100.00 (280)	100.00 (595)	100.00 (1217)	100.00 (56742)	100.00 (69531)	100.00 (122428)	100.00 (259343)	100.00 (18405)	100.00 (20681)	100.00 (29158)	100.00 (56953)
Share of export-oriented sectors as % of all industry			96.85	98.41	66.24	60.30	58.90	68.34	38.26	34.93	41.48	48.83	50.41	40.86	42.12	44.95
All industry in unit **			74636	159695	318	464	1009	1781	148316	199039	295184	531074	36511	50613	69227	126711

Note: **Unit for production workers in '000; others are in million taka.

1. Export-oriented sectors are identified based on export as per cent of total sales for 2000 (the latest year's export). A sector is identified as export-oriented if the export as per cent of total sales exceeds 20 per cent. A total of 21 sectors (12 per cent of total 179 4-digit industry sectors) satisfy this criterion.
2. PR, P-1, P-2 & P-3 means Pre-reform (1978-1980), Phase 1 (1981-1990), Phase 2 (1991-1995) & Phase 3 (1996-2000) respectively.
3. The industries are presented according to the ascending order of capital-labour ratio.

Source: Author's estimation based on CMI reports, various years.

Annex Table A-5.5

Labour Productivity Index (Base: 1991=100) at 2-Digit Level CMI Data

Industry Description (code)	*1991*	*1992*	*1993*	*1994*	*1996*	*1998*	*2000*
Manufacture of food beverages and tobacco (31)	100 (624.35)	100.403 (626.87)	104.07 (649.73)	165.25 (1031.73)	182.46 (1139.19)	174.41 (1088.93)	169.43 (1057.81)
Textile wearing apparel and leather industries (32)	100 (149.01)	102.403 (152.59)	106.12 (158.12)	145.47 (216.76)	139.7 (208.16)	113.35 (168.897)	116.58 (173.71)
Manufacture of wood & wood products, including furniture (33)	100 (137.69)	101.214 (139.36)	178.57 (245.87)	133.67 (184.04)	400.03 (550.78)	111.77 (153.88)	191.32 (263.42)
Manufacture of paper & paper products (34)	100 (491.25)	100.928 (495.81)	87.973 (432.17)	85.484 (419.94)	132.6 (651.40)	107.27 (526.98)	90.291 (443.56)
Manufacture of chemicals & chemical petroleum & rubber (35)	100 (1015.6)	102.465 (1040.6)	117.97 (1198.1)	85.672 (870.05)	126.75 (1287.21)	129.34 (1313.49)	118.06 (1199.02)
Manufacture of non-metallic mineral products (36)	100 (172.84)	126.188 (218.1)	136.27 (235.53)	203.17 (351.16)	137.69 (237.97)	221.63 (383.06)	95.45 (164.97)
Basic metal industries (37)	100 (1079.7)	116.155 (1254.2)	123.06 (1328.7)	180.67 (1950.75)	143.9 (1553.69)	171.19 (1848.38)	145.86 (1574.85)
Manufacture of fabricated metal production machinery & equipments (38)	100 (425.57)	94.2971 (401.3)	118.14 (502.75)	105.49 (448.93)	125.1 (532.39)	169.9 (723.06)	176.88 (752.75)
Others (39)	100 (167.71)	92.6822 (155.43)	146.86 (246.29)	548.96 (920.64)	768.75 (1289.24)	695.39 (1166.20)	469.38 (787.17)
All mfg. industries	100 (265.29)	99.3447 (263.56)	105.02 (278.6)	136.08 (361.00)	128.44 (340.73)	108.92 (288.95)	103.84 (275.48)

Note: Figures in the parentheses indicate '000 taka/year.

Source: Authors' estimation based on CMI data, various issues.

Annex Table A-5.6

Real Wage Index (Base: 1991=100) at 2-Digit Level CMI Data

Industry Description (code)	*1991*	*1992*	*1993*	*1994*	*1996*	*1998*	*2000*
Manufacture of food beverages and tobacco (31)	100 (18.66)	117.34 (21.89)	90.15 (16.82)	189.65 (35.38)	208.84 (38.96)	58.25 (10.87)	250.22 (46.68)
Textile wearing apparel and leather industries (32)	100 (24.99)	57.37 (14.34)	78.67 (19.66)	82.83 (20.70)	65.67 (16.41)	167.49 (41.86)	172.14 (43.02)
Manufacture of wood & wood products, including furniture (33)	100 (12.11)	351.76 (42.61)	202.50 (24.53)	108.19 (13.11)	176.13 (21.34)	172.93 (20.95)	146.31 (17.72)
Manufacture of paper & paper products (34)	100 (35.54)	131.62 (46.78)	42.23 (15.01)	116.21 (41.31)	70.66 (25.11)	85.74 (30.47)	61.71 (21.93)
Manufacture of chemicals & chemical petroleum & rubber (35)	100 (18.51)	117.09 (21.67)	208.09 (38.51)	180.92 (33.48)	59.84 (11.07)	137.12 (25.37)	71.93 (13.31)
Manufacture of non-metallic mineral products (36)	100 (15.86)	172.31 (27.33)	144.44 (22.91)	168.75 (26.77)	216.43 (34.33)	249.54 (39.58)	79.69 (12.64)
Basic metal industries (37)	100 (36.26)	75.26 (27.29)	45.78 (16.60)	51.68 (18.74)	40.02 (14.51)	53.65 (19.46)	161.41 (58.53)
Manufacture of fabricated metal production machinery & equipments (38)	100 (33.80)	61.13 (20.66)	85.10 (28.76)	64.67 (21.86)	38.78 (13.11)	55.21 (18.66)	50.89 (17.20)
Others (39)	100 (12.99)	242.27 (31.46)	235.54 (30.59)	99.28 (12.89)	204.94 (26.62)	203.20 (26.39)	115.88 (15.05)
All mfg. industries	100 (22.49)	97.35 (21.89)	98.83 (22.22)	86.26 (19.40)	106.15 (23.87)	72.88 (16.39)	79.63 (17.91)

Note: Figures in the parenthesis indicate wage per employee in '000 tk/year.

Source: Authors' estimation based on CMI data, various issues.

6 Impact of Trade Liberalisation on Labour Demand and Growth in the Manufacturing Industries of Bangladesh

An Econometric Investigation

Establishing causal relationship between trade liberalisation and employment is critical to understanding the impact of the former on the latter. However, this is not an easy task. In the previous chapter, an attempt has been made to examine the movements in the labour market, at sectoral and disaggregate levels in relation to the various phases of trade liberalisation. This chapter will investigate the impact of trade liberalisation on the growth in output, labour demand elasticities and worker's wage in the manufacturing industries in Bangladesh by applying advanced econometric tools. A Cobb-Douglas type production function is estimated for Bangladesh's manufacturing sector with and without trade liberalisation variable[1] using panel data compiled from CMI reports covering the time period between 1991 and 2000. Following this, a theoretically consistent labour demand function is estimated and the impact of trade liberalisation on labour demand elasticities is examined.

Before proceeding to the modelling exercise, the correlation coefficients are looked at for a choice of variables to get an idea about the bivariate relationship among them. The results are presented in Annex A-6.1 and Annex Table A-6.1.

1. For the description of variables (used in various regression function), please see Annex Table A-6.2.

6.1 Estimation of Production Function in the Manufacturing Industries of Bangladesh

The analysis begins by employing 'production function' approach to examine the relationship between trade liberalisation and output growth in the context of the manufacturing industries of Bangladesh. The advantages with this approach are two-fold (Raihan, 2007): (1) with this approach, relevant factors of production (such as, capital and labour) can be controlled for, which helps to examine the impact of trade liberalisation measures on growth; (2) the production function approach identifies the sources of growth, which are central to the neoclassical model of Solow (1956). The Solow model considers three sources of growth: the accumulation of physical capital, the growth of labour force and the growth of total factor productivity.

It should, however, be mentioned here that the neoclassical production function approach to economic analysis has some drawbacks as well. There are concerns about many of the assumptions of the neoclassical production function, i.e., perfect competition and perfect mobility of factors of production, in the context of both the developing and developed countries. Furthermore, the use of an aggregate production function assumes that the sectors are symmetrical. Despite these limitations, the production function approach to examine trade liberalisation and growth relationship has been popular among researchers (Raihan, 2007).

6.1.1 The Model

With respect to the functional form of the production process, several functions, such as a Cobb-Douglas, CES or translog, can be assumed to estimate the production function. However, in the empirical literature the Cobb-Douglas functional form has been most extensively used in the production function analysis, thanks to the flexibility it provides in dealing with data imperfections (Tybout 1990). Moreover, a Cobb-Douglas functional form can afford the maximum analytical and mathematical flexibility. Besides, it fits the Bangladesh manufacturing data reasonably well (Raihan, 2007; Krishna and Sahota, 1991). The analysis of this study starts with a Cobb-Douglas production function to estimate the basic Solow-type sources of growth, and then incorporates various trade liberalisation measures in the original model to examine whether the variables representing trade liberalisation turn out to be statistically significant or not.

In a panel framework, the benchmark model is specified and defined for each sector and each year as follows:

$$Y_{it} = A_{it} K_{it}^{\alpha_i} L_{it}^{\beta_i} \tag{6.1}$$

where, Y_{it} is real output for sector i in time period t, K_{it} is the capital stock and L_{it} is the number of workers. α and β are the average shares of capital and labour respectively in total value-added.

It is to be mentioned here that A_{it} is the technology specified in the model which includes a possible intercept (θ_i), trend term *(t)*, two dummy variables one for labour-intensive sectors *(l_int)* and another for export oriented sectors (*X_or*) and an interaction term between trend and capital:

$$A_{it} = \exp(\theta_i + \gamma_{it} \ln t + \lambda_{it} \ln t \ln k_{it} + \delta_1 l\_\text{int} + \delta_2 X\_or + \varepsilon_{it}) \tag{6.2}$$

It has already been mentioned earlier that labour-intensive industries and export-oriented industries are expected to display distinctive production process as opposed to capital-intensive and import-competing industries. Taking note of this, two dummy variables have been introduced in the model. Taking the natural log of (4.1) and substituting the value of A_{it} forms the following model:

$$\begin{aligned} \ln Y_{it} &= \theta_i + \alpha_i \ln k_{it} + \beta_i \ln L_{it} + \gamma_{it} \ln t + \lambda_{it} \ln t \ln k_{it} \\ &\quad + \delta_1 l\_or + \delta_2 X\_or + \varepsilon_{it} \end{aligned} \tag{6.3}$$

In a panel framework, θ_i is considered as the unobservable industry fixed effect and ε_{it} is the idiosyncratic error satisfying strict exogeneity assumption; i.e., $E(\varepsilon_{it} | X_i, \theta_i) = 0$ where X_i represents the set of explanatory variables included in the model.

To verify the link between trade liberalisation and growth, the model (4.1) could be extended further by including some measures of trade liberalisation. The underlying hypothesis of such inclusion is that apart from the Solow-type basic sources of growth, the measures of trade liberalisation may have a significant influence on growth. Such inclusion of trade liberalisation variables results in:

$$Y_{it} = A_{it} k_{it}^{\alpha_i} L_{it}^{\beta_{it}} g_{it}^{\eta_i} \quad (6.4)$$

where g represents some measures of trade liberalisation, such as the tariff rate or the import-penetration ratio. With the help of (4.2) and the natural logarithm of (4.4), the final equation to be estimated is:

$$\ln Y_{it} = \theta_i + \alpha_i \ln k_{it} + \beta_i \ln L_{it} + \gamma_{it} \ln t + \lambda_{it} \ln t \ln k_{it} + \delta_1 l\_\text{int} + \delta_2 X\_\text{int} + \eta_i \ln g_{it} + \varepsilon_{it} \quad (6.5)$$

As Harrison (1996) points out, inclusion of trade liberalisation measures in the production function is consequently a test of their impact on technological change—growth in output due to technological change after controlling for other factors of production. This is also the idea behind 'endogenous growth theory' that establishes a link between trade and growth through what is called the 'technological spill over effects' (Grossman and Helpman, 1991).

Raihan (2007) estimated the long run elasticities of labour and capital for the production function of Bangladesh's manufacturing sector to be 0.602 and 0.308 respectively using the unrestricted error correction mechanism (UECM) with panel industry data set according to ISIC 3-digit code. However, using five different indicators of trade liberalisation the study could not find any significant positive contribution of trade liberalisation on the growth of the value added.[2] Rather, the study found significant negative impact of trade liberalisation on output growth in two cases; for example a 1 per cent increase in sectoral import penetration ratio leads to a 0.40 reduction in the output growth.

6.1.2 Variables and Data Description

Data requirement for estimation of a benchmark Cobb-Douglas production in a panel framework, as specified in (6.3) or (6.5) function is met by what is available in the census of manufacturing industries reports.

2. Five indicators are import penetration of consumer goods, implicit nomial tariff rate, sectoral import penetration ratio, sectoral export orientation ratio and a dummy variable which defines 1992 as the demarcation point of pre- and post-reform. However, the first two indicators are not available at the disaggregated industry data and assumed that they remain constant across industries but vary over time.

The outlier corrected industry data at 4-digit level BSIC code (converted to real by using GDP deflator series) covering the time period of 7 years (1991 to 2000) has been used for the purpose of the present analysis. In the literature on the estimation of the production function, both value added and gross output are used in tandem. However, in the Bangladesh Census of Manufacturing Industries (CMIs), estimates of intermediate inputs are not available consistently over the period of analysis. Consequently, this study considers real value-added as the indicator of real output in the estimation of the production function. The number of production workers was taken to as a proxy for the number of labour, while the capital stock has been estimated by using perpetual inventory approach as follows:

$$K(t) = k(0) + \sum I(t) - \delta . k(0)$$

where, $K(t)$ and $k(0)$ are the capital stock at time t and at initial period, $I(t)$ is the investment and δ series is the rate of depreciation. More specifically, the real capital stock used in this analysis is a derived variable using the variables fixed asset, fixed capital formation and depreciation, reported in the CMI (all converted in real terms).

To quantify the extent of globalisation, sectoral trade (export, import and trade) to value added ratio and tariff data (custom duty rate and tariff incidence rate separately) have been compiled from UN Comtrade and Bangladesh Tariff Commission respectively.[3] It is to be mentioned here that to avoid missing values while taking natural logarithm, zero trade and tariff values were replaced by a very small quantity.

Finally, the 21 and 65 sectors at 4-digit BSIC code, as identified in Chapter 5, were considered for defining the export-oriented and labour intensive dummies. (See Table 6.2 for variable description.)

6.1.3 Estimation Strategy

A total of 12 specifications of the production function have been estimated with the application of fixed effect estimation method for panel

3. For sectoral trade and tariff data, the BSIC-ISIC-HS mapping has been used, please see Chapter 3 for details.

data. In this analysis (and also for labour demand function), static panel methods were applied instead of dynamic panel models mainly due to unavailability of trade liberalisation data. At first, the bench mark model (i.e., the model without trade and tariff variable) as specified in (6.3) has been estimated. Following this, the benchmark panel model has been augmented by two types of tariff data and three types of trade-value added ratio data separately to give five alternative specifications. Finally, six other models have been estimated by incorporating one tariff and one trade variable at a time. Such an exercise was expected to provide an opportunity to examine the robustness of the parameter estimates.

It is to be mentioned here that the estimated elasticities should be considered as long-run elasticities since a static panel framework has been considered and no lagged term of the variables were entered into the model.

Though it is well accepted and a usual practice in empirical literature to use fixed effect technique for industry level panel data, a number of diagnostic tests have also been performed before applying this technique. For each model, Hausman's (1978) specification test to choose between fixed effect and random effect models has been performed. Besides, Breusch and Pagan's (1980) Lagrangian multiplier (LM) test for random effects has been performed.

6.1.4 Discussion on Results

Table 6.1 displays the estimation results of the above-mentioned 12 specifications of the production function with a number of diagnostic tests. The values of overall F-statistic show that all models are statistically significant at 1 per cent level of significance. Results of both Hausman and LM tests suggest the use of fixed effect model. The estimates of coefficients associated with labour and capital are robust, highly significant and, most importantly, are of expected sign. Elasticities of labour with respect to output lie within the range of 0.539 to 0.611 while elasticities of capital lie within the range of 0.183 to 0.208. The sum of labour and capital elasticities lies between 0.735 and 0.816. Model 7 (i.e., specification with sectoral trade to VA ratio as liberalisation variable) yields the highest labour elasticity indicating that 1 per cent increase in labour leads to 0.611 per cent increase in the value-added. In all the models, no significant impact of time trend over the production process has been found. Both the dummy

Table 6.1

Estimation of the Cobb-Douglas Production Function for Bangladesh's Manufacturing Sector

	Without Tariff and Trade	*Tariff Only*		*Trade Only*			*With Trade and Tariff*					
		Custom Duty (CD)	*Tariff Incidence (TI)*	*Export (X)*	*Import (M)*	*Trade*	*X & CD*	*X & TI*	*M & CD*	*M & TI*	*Trade & CD*	*Trade & TI*
Model	*Model1*	*Model2*	*Model3*	*Model4*	*Model5*	*Model6*	*Model7*	*Model8*	*Model9*	*Model10*	*Model11*	*Model12*
Labour	0.605*	0.608*	0.606*	0.607*	0.557*	0.539*	0.611*	0.607*	0.560*	0.557*	0.542*	0.539*
Capital	0.207*	0.206*	0.208*	0.206*	0.185*	0.196*	0.205*	0.207*	0.183*	0.186*	0.195*	0.197*
Custom duty		-0.024					-0.029		-0.043		-0.030	
Tariff incidence			0.030							0.030		0.032
Export orientation				-0.021			-0.022	-0.020				
Import penetration					-0.143*				-0.144*	-0.143*		
Trade orientation						-0.140*					-0.140*	-0.140*
Trend	-0.070	-0.080	-0.045	-0.048	0.078	0.093	-0.060	-0.028	0.054	0.103	0.079	0.120
Trend*capital	0.019	0.019	0.019	0.018	0.014	0.014	0.017	0.017	0.013	0.013	0.013	0.014
Constant	4.438*	4.550*	4.272*	4.500*	5.577*	5.534*	4.643*	4.366*	5.795*	5.412*	5.678*	5.357*
Sample size	1126	1124	1126	1126	1126	1126	1124	1126	1124	1126	1124	1126
Correlation	0.593	0.590	0.594	0.596	0.583	0.612	0.593	0.596	0.582	0.581	0.611	0.610
F-value	184.260*	147.400*	147.340*	147.990*	175.820*	179.040*	123.430*	123.240*	146.930*	146.440*	149.350*	149.130*
F test (for fixed)	5.810*	5.790*	5.740*	5.730*	6.330*	6.300*	5.720*	5.670*	6.310*	6.300*	6.280*	6.270*
χ^2 (bplm)	469.050*	468.760*	459.170*	458.870*	485.150*	488.120*	458.880*	450.230*	485.570*	481.280*	487.930*	484.640*
χ^2 (haus)	59.630*	60.150*	62.100*	61.760*	96.480*	91.890*	62.320*	64.560*	93.170*	95.790*	88.970*	91.700*

Note: 1. *,**, *** indicate significance at 1 per cent, 5 per cent, 10 per cent level respectively.

2. All variables are measured in natural logarithm. Labour and capital, denote production worker and capital stock respectively.

3. Both labour-intensive and export-orientation dummies are dropped.

4. Correlation indicates the correlation between unobserved effects with explanatory variables. χ^2 (bplm) and χ^2 (haus) represent the test statistics value of Hausman's Specification Test and Breusch and Pagan Lagrangrian Multiplier Test for Random Effects.

variables for labour-intensive and export-oriented industries were dropped. The implication of this result is that the CMI data under investigation may not capture any distinctive characteristics for labour-intensive (export oriented) industries as opposed to non-labour intensive (non-export oriented) industries.

The key variables under investigation represent impact of trade liberalisation on production function. Interestingly, none of the coefficients associated with tariff variables (namely custom duty and tariff incidence) and sectoral export to VA ratio have turned out to be significant. On the other hand, two other trade variables i.e., sectoral import to VA ratio and sectoral trade to VA ratio have significant and robust negative impact on growth of output. For example, a 1 per cent increase in sectoral import penetration results in 0.143 per cent decrease in growth in output according to both Model 5 and Model 10, and 0.144 per cent decrease in growth in output according to Model 9. In other words, trade liberalisation has not had any significant positive impact on overall manufacturing production; rather, sectoral trade and import orientation appears to have adverse impact on the production process. The findings with respect to import penetration are consistent with the findings of Raihan (2007).

6.2 Labour Demand Function

6.2.1 Background and Literature Review

In most of the recent studies on the effects of increasing integration with the global economy on the labour market, the major thrust has been on evaluation of the impact of trade on wages and employment. However, there could be various channels through which globalisation could influence the labour market. One of these is the effect on labour demand elasticity. Leamer (1995) and Wood (1995) have suggested that demand for labour is more elastic for an open economy compared to when the economy is closed. Rodrik (1997) argued that a higher responsiveness of labour demand to changes in labour prices might be a direct consequence of international economic integration "regardless of economic structure and the identity of the trade partners" (Rodrik, 1997: 26). On the other hand, Panagariya (1999) theoretically showed that Rodrik's assumption of positive effect of globalisation on labour demand elasticity is not a general result. As a consequence, the validity of that relationship has to be determined empirically.

One of the major thrusts of this study has been to empirically estimate the labour demand elasticities of Bangladesh for the manufacturing sector. There are a number of empirical studies, both country-specific as well as cross-country, on estimating the labour demand function and also to test Rodrik's hypothesis mentioned above. Using sophisticated econometrics, Bruno *et al.* (2004) carried out a comparative study of the OECD countries, by estimating a single stage labour demand function for balanced panel data and found evidence in favour of Rodrik's conjecture only in the case of France and the UK. Slaughter (2001) estimated labour demand function in two stages, taking the industry-wise panel data from 1961 to 1991 for the United States, which provides mixed support for Rodrik's views. Krishna *et al.* (2001) used the wage variable interacted with a liberalisation dummy and failed to establish that labour demand is more elastic in response to trade liberalisation in the case of Turkey. In a recent cross-country analysis focusing on the Latin American LDCs, Fajnzylber and Maloney (2001) found no consistent pattern for the impact of trade liberalisation. However, using a dynamic labour demand function for UK, Greenaway *et al.* (1999) found that increases in trade volumes, both in terms of imports and exports, cause reductions in the level of derived labour demand justifying the view that trade liberalisation brings efficiency. They also analysed the impact of trade changes on the slope of the derived labour demand introducing a wage-trade interaction term and found positive but insignificant impact of trade volumes on the labour demand elasticity. For France, Jean (2000) found significant effect of trade liberalisation on labour demand elasticities by applying a different methodology and focusing on the intersectoral dimension of the scale effect. In a recent study, Bruno *et al.* (2005) found significantly negative labour demand elasticities, both short-run and long-run and Rodrik's supposition is decidedly rejected for various estimators used, which confirms the results for Italy in Bruno *et al.* (2004). The direct effect of globalisation on labour demand is not found to be significant; rather, robust evidence was found in favour of output generated external economies.

Trade could induce an increase in labour demand elasticity through scale effect caused by the increased competition on the output market and/ or via a substitution effect generated by expansion of firm's production possibility set to utilise additional inputs or by increasing productivity (Hamermesh, 1993).[4] With reference to the substitution effect, greater international openness enables easier access of firms to intermediate products and capital equipment, both domestically and abroad, and possibility to directly substitute domestic factors of production by foreign factors. In addition, trade may be a vehicle for technological spillovers, through both the import of goods embodying foreign knowledge and/or the acquisition of useful information that would be otherwise costly to obtain (Coe and Helpman, 1995). All these factors contribute to expand the business and technological opportunities of the firm, enhancing not only the substitutability among factors of production, but also technical efficiency. The focus of this exercise is centred on the latter channel of transmission.

The experience of substantial changes in the trade regime in Bangladesh over the decade of 90s might be thought of as an appropriate context to investigate the theoretical link between openness and labour demand elasticities. A labour demand equation is obtained from the solution of a firm's cost minimisation problem following the general framework developed by Bruno *et al.* (2005), with minor modification for allowing to include the tariff variable along with the trade variable as indicator of the extent of globalisation. This theoretically consistent labour demand is estimated for Bangladesh using a 7-year panel data from published CMI data over the period 1991-2000.

In the present econometric exercise a major focus has been to measure constant output own-price labour demand elasticity evaluating the impact of globalisation. Globalisation could possibly have a direct effect on the labour productivity which act as labour demand shifter in the domestic employment market (Greenaway *et al.*, 1999) as well as an indirect impact

4. It is well known that total own-price labour demand elasticity is composed of two parts: 1) the scale effect, which explains the employment variation due to the wage-induced change in the demanded output; and 2) the substitution effect, which explains the employment variation due to substitution toward other inputs for constant output.

on labour market through influencing the real wages of workers. To estimate this indirect impact, which captures the change in responsiveness of labour demand to changes in labour prices as mentioned by Rodrik (1997), an interaction term of wage with globalisation variable was introduced in this model. The specification adopted in this analysis allows to estimate and to test both effects in a unique estimation run by treating the globalisation variable as a shifter for both the labour demand equation and the labour elasticity. Moreover, to estimate a globalisation effect free of technical change bias, we also included a time trend.

6.2.2 The Model

The underpinning theoretical model on which the labour demand equation has been postulated is rather simple, with the feature of producing labour demand elasticities in one step, thereby avoiding the econometric difficulties of two-step procedures. The existence of a representative firm for each sector of the economy is assumed which utilises two domestic production inputs, domestic labour l and capital k producing output y, with w and p being the compensations for l and k, respectively. Without loss of generality, p can be considered as unity (i.e., normalising by p) and then w represents the relative compensation for domestic labour. The market for production factors is perfectly competitive, whereas no assumption is made on the form of the output market.

Trade activity g_1 and tariff rate g_2 are key variables of the model to proxy the extent of globalisation. Both g_1 and g_2 are allowed to play a role to influence the domestic production possibility set of the firm. However, it is important to note that, since higher tariff rates means lower trade, g_1 and g_2 should have negative correlation, if there is any.[5] Therefore, under the assumption of positive impact of globalisation, it is expected for g_1 to have positive relationship on labour demand while g_2 should have negative coefficient. International trade might exert an impact on the production possibility set through availability of inputs and new production techniques that were previously not accessible or unknown to the firm. Also, trade might act as productivity enhancer of the existing inputs by bringing new

5. In our case, both tariff variables (custom duty and tariff incidence) have no significant relationship with exports but have negative and significant relationship with import for all sectors. See Annex A-6.1 for further understanding.

intangible inputs, such as new foreign knowledge and the acquisition of useful information, or simply by acting as a yardstick of performance for the domestic labour, inducing more effort in the workplace. Overall, these forces may influence labour's own price elasticity, as well as bring about a direct effect on labour demand, with *g* acting as a demand shifter.

The sequencing of firm decisions is as follows. At first stage, upon observing relative price *w* and a number of other relevant exogenous variables *x*, the firm chooses the degree of openness to the global market *g* and the level of production *y*. Nothing has been assumed about the cost of globalisation and about the optimality of output and openness decisions. Secondly, the firm experiences an idiosyncratic productivity shock ε, such that $E(\varepsilon|w,x)=0$, and chooses the domestic factor demands to minimise its cost $C = wl + k$ such that $E(\varepsilon|w,x,g,y)=0$

Under these assumptions, the domestic labour demand is specified as follows:

$$ln(l_{it}) = \alpha_i + \beta_w ln(w_{it}) + \beta_y ln(y_{it}) + \beta_{g(1)} ln(g_{(1)it}) + \beta_{g(2)} ln(g_{(2)it}) + \beta_{wg(1)} ln(w_{it}) * ln(g_{(1)it})$$
$$+ \beta_{wg(2)} ln(w_{it}) * ln(g_{(2)it}) + \beta_t * ln(t_{it}) + \beta_{wt} ln(w_{it}) * ln(t_{it}) + \delta_1 l\_int + \delta_2 x\_or + \varepsilon_{it}$$

$$i = 1,2,\cdots,N \quad and\ t = 1,2\cdots,T \qquad (6.6)$$

Where,

α_i= sector specific fixed effect

w_i= average wage per production worker for i-th sector at time *t*

y_{it} = value added at factor cost for i-th sector at time *t*

$g_{(1)it}$= Globalisation indicator 1 (ratio of sectoral trade data (trade, export or import) to sectoral value addition)

$g_{(2)it}$= Globalisation indicator 2 sectoral tariff data (both custom duty and tariff incidence separately)

l_int = dummy for labour-intensive sectors

x_or = dummy for export-orientation

t_{it}= time trend

And $\ln(w_{it})*\ln(g_{(1)it})$; $\ln(w_{it})*\ln(g_{(2)it})$ *and* $\ln(w_{it})*\ln(t_{it})$ are the interaction terms wage variable with trade, tariff and trend respectively.

$\beta_w, \beta_y, \beta_{g(1)}, \beta_{g(2)}, \beta_{wg(1)}, \beta_{wg(2)}, \beta_t, \beta_{wt}, \delta_1, \delta_2$ are unknown parameters. β_{wg1}, β_{wg2} and β_{wt} measures the impact of g_1, g_2 and t on the relative wage elasticity of the labour demand function. According to neoclassical model, β_w must be always negative (i.e., inverse relationship between wage and employment) while β_y should be always positive as more output requires more labour. An equivalent form of (6.6) is given by:

$$\ln(l_{it}) = \alpha_i + [\beta_w + \beta_{wg(i)} \ln g_{(1)it} + \beta_{wg2} \ln g_{(2)it} + \beta_{wt} \ln t_{it}] \ln w_{it} + \beta_y \ln y_{it} + \beta_{g(1)it} \ln g_{(1)it} + \beta_{g2} \ln g_{(2)it} + \beta_x \ln x_{it} + \varepsilon_{it} \tag{6.7}$$

where x_{it} includes time trend t_{it} and two dummies: one for labour intensive sector and another for export-oriented sector.

It is to be mentioned that since the production process would continue even with zero trade or zero tariff value, missing value and zero values of trade variables are replaced by 1 such that $g_i \geq 1 \; \forall \; i,t$.

Now relative wage elasticity of the labour demand function,

$$\begin{aligned} \in_{lw} &= \frac{\partial \ln l}{\partial \ln w} \\ &= \beta_w + \beta_{wg(1)} * \ln(g_{(1)}) + \beta_{wg(2)} * \ln(g_{(2)}) + \beta_{wt} * t \end{aligned} \tag{6.8}$$

For the labour demand function to be theoretically consistent with the cost minimising behaviour and a non-decreasing production function, $\in_{lw}$ must lie between [-1,0].

The economic interpretation of β_g parallels that β_w in that β_g is the intercept of the labour elasticity with respect to g,

$$\in_{lg(j)} = \frac{\delta l}{\delta g_{(j)}} = \beta_{g(j)} + \beta_{wg} \ln w \tag{6.9}$$

In other words, β_g measures the responsiveness of labour demand to g at w=1, that is when this economic rate of substitution is 1.

From the derivation of production function presented in the Annex A-6.3, the elasticity of A with respect to or generalised technical efficiency effect:

$$\in_{Ag} = -\frac{\beta_g}{\beta_y} + \frac{\beta_{wg}}{\beta_y} \left[\ln\left(-\frac{\in_{lw}}{\in_{lw}+1}\right) + \frac{1}{\in_{lw}+1} \right], \quad \in_{Ag} \in \Re \qquad (6.10)$$

It is to be noted here that *if* $\beta_{wg}=0$, then $\in_{Ag} = -\frac{\beta_g}{\beta_y}$ is a constant parameter with g acting on the isoquant mapping as Hicks-neutral technical change, as assumed by Greenaway *et al.* (1999).[6] Unlike β_g and β_y, this formulation of generalised technical efficiency effect puts no theoretical restriction on the sign of $\in_{Ag}$ and the presence of β_g ensures enough flexibility to cover all possible relevant economic instances (Bruno *et al.*, 2005).

Before proceeding further, one point needs to be made clear that a negative coefficient of β_{wg}, as assumed by Rodrik, implies positive impact of g on labour demand elasticities (as both β_w and β_{wg} would act then in the same direction). In this formulation, g is implicitly assumed to be an increasing function of the extent of globalisation. In this case, globalisation indicators based on trade data indeed satisfy this inherent supposition; however, globalisation indicators based on tariff data is a decreasing function of trade liberalisation meaning decrease in tariff indicates more openness. In turn this implies that a negative coefficient of tariff-wage interaction term leads to the conclusion that protection has positive impact on labour demand elasticity. Therefore, in this case to hold Rodrik's conjecture of positive impact of globalisation, β_{wg} should reverse its sign (i.e., $\beta_{wg} > 0$), as formulated in usual cases. This distinction should be borne in mind while interpreting the coefficients and elasticity.

---

6. Neutral technological change refers to the behaviour of technological change in models. A technological innovation is Hicks-neutral (following John Hicks (1932)), Harrod-neutral (following Hicks) or Solow-neutral if the ratio of capital's marginal product to labour's marginal product is unchanged, labour-augmenting (i.e., helps labour) or capital-augmenting (i.e., helps capital) for a given capital to labour ratio.

6.2.3 Data, Variables and Estimation Strategy

To estimate the panel labour demand function as specified (4.6) or (4.7), the required data is collected from various CMI reports. Like the production function, in this case also, outlier corrected real data on production worker, per production worker average wage, and value added at factor cost over the period of 7 years (1991 to 2000) has been used. The definitions of the five variables (two tariff based and three trade based) denoting the degree of trade liberalisation and two dummies for labour intensive and export-oriented industries, remain same as mentioned earlier for production function (see Annex Table A-6.2 for the variable description). To avoid missing values while taking natural logarithm, zero trade and tariff values are replaced by a very small quantity.

Since the model specified has the feature to provide elasticities using a single stage estimation procedure, the estimation strategy for the labour demand function is essentially same as the production function due to the same data structure (e.g. 12 specifications of the model, static fixed effect estimation method, diagnostic test etc.).

6.2.4 Analysis of the Regression Results

Table 6.2 presents the estimation results of the above-mentioned 12 specifications of the labour demand function with a number of diagnostic tests. The F-test for overall model confirms the statistical significance of the model while both Hausman and LM tests provide evidence for applying fixed effect technique. The estimates of the coefficients associated with value-added at factor are consistent with theory, robust and highly significant (within the range of 0.50 to 0.52). However, another core variable wage per employee appears to be significant in case of the benchmark model (model 1) and all three trade only models (model 4, model 5 and model 6), all of which are negative (within the range of -0.17 to -0.30) satisfying the theoretical restrictions. The estimates of the coefficients associated with time trend variable is positive and significant; and robust to the model specification indicating that during the time period under consideration, the demand for manufacturing labour has grown in general in Bangladesh. The robust impact of time trend on relative wage (estimated by the trend-wage interaction) is found to be negative which means an increase in real wage over the years yields lower demand for

Table 6.2

Estimation of Labour Demand Function for Manufacturing Sector of Bangladesh

	Without Tariff and Trade	*Tariff Only*		*Trade Only*			*With Trade and Tariff*					
		Custom Duty (CD)	*Tariff Incidence (TI)*	*Export (X)*	*Import (M)*	*Trade*	*X & CD*	*X & TI*	*M & CD*	*M & TI*	*Trade & CD*	*Trade & TI*
Variables	*Model1*	*Model2*	*Model3*	*Model4*	*Model5*	*Model6*	*Model7*	*Model8*	*Model9*	*Model10*	*Model11*	*Model12*
Output	0.519*	0.520*	0.519*	0.517*	0.516*	0.508*	0.518*	0.517*	0.518*	0.515*	0.509*	0.508*
Wage	-0.174**	0.429	0.217	-0.288*	-0.263*	-0.301*	0.290	0.062	0.294	0.067	0.251	0.0425
Custom duty		0.371**					0.343***		0.328***		0.328***	
Tariff incidence			0.204					0.179		0.164		0.172
Export orientation				-0.077***			-0.061	-0.074				
Import penetration					-0.066***				-0.050	-0.061		
Trade orientation						-0.099***					-0.090**	-0.097*
Custom duty*wage		-0.123**					-0.114***		-0.110***		-0.110***	
Tariff incidence*wage			-0.076					-0.067		-0.063		-0.066
Export*wage				0.028***			0.023	0.027				
Import*wage					0.022				0.017	0.020		
Trade*wage						0.031**					0.029**	0.030**
Trend	0.233***	0.531***	0.368**	0.252***	0.283**	0.322**	0.524***	0.370**	0.535*	0.388**	0.578*	0.435**
Trend*wage	-0.077***	0.177*	-0.127**	-0.084***	-0.092**	-0.102**	-0.175***	-0.128**	-0.177*	-0.132**	-0.188*	-0.148**
Constant	1.279*	-0.544	0.225	1.620*	1.571*	1.787*	-0.136	0.684	-0.114	0.711	0.133	0.894
Sample size	1129	1127	1129	1129	1129	1129	1127	1129	1127	1129	1127	1129
Correlation	0.721	0.717	0.7208	0.725	0.723	0.727	0.722	0.725	0.719	0.723	0.723	0.727
F-value	186.52*	125.62*	124.56*	125.24*	124.96*	126.17*	94.67*	94.02*	94.40*	93.78*	95.30*	94.72*
F-test (fixed effect)	10.680*	10.740*	10.63*	10.650*	10.630*	10.630*	10.690*	10.600*	10.650*	10.590*	10.65*	10.59*
χ^2 (bplm)	743.99*	727.30*	727.3*	723.35*	736.56*	732.06*	729.74*	713.55*	741.79*	728.72*	735.85*	725.09*
χ^2 (haus)	165.96*		111.43*	na	129.05*	132.42*	244.59*	na	116.83*	120.35*	112.07*	116.16*

Note: 1. *,**, *** indicate significance at 1 per cent, 5 per cent, 10 per cent level respectively.

2. Output and wage represent value added factor cost and production worker wage respectively.

3. Both labour-intensive and export-orientation dummies are dropped.

4. Correlation indicates the correlation between unobserved effects with explanatory variables. χ^2 (bplm) and χ^2 (haus) represent the test statistics value of Hausman's Specification Test and Breusch and Pagan Lagrangrian Multiplier Test for Random Effects.

labour. Similar to the production function, here also the labour-intensive and export-orientation dummies have been dropped.

Table 6.3

Direct and Indirect Impact of Trade Liberalisation on Labour Demand and Relative Wage

Specifications	*Direct Impact of Trade Liberalisation on Labour Demand*		*Impact of Trade Liberalisation on Relative Wage*	
	Tariff Variable	*Trade Variable*	*Tariff-wage Interaction(β_{wg1})*	*Trade-wage Interaction(β_{wg2})*
Model 2 (custom duty only)	0.371		-0.123	
Model 4 (export only)		-0.077		0.028
Model 5 (import only)		-0.066		0.022
Model 6 (trade only)		-0.099		0.031
Model 7 (export and custom duty)	0.343		-0.114	
Model 9 (import and custom duty)	0.328		-0.110	
Model 11(trade and custom duty)	0.328	-0.090	-0.110	0.028
Model 12 (trade and tariff incidence)		-0.097		0.030

Note: Extracted from Table 6.2.

To examine the impact of trade liberalisation on labour demand elasticities, major interest lies in the interpretations of the coefficients associated with globalisation variables (which give the estimate of the direct impact of trade liberalisation) and their interaction terms with wage to capture the indirect impact of trade liberalisation which may influence labour own price elasticity. Table 6.3 shows direct and indirect impacts of globalisation on labour demand and relative wage respectively.

First, the direct impact of trade liberalisation on labour demand as a demand shifter is discussed. Significant positive direct employment impact of more restricted trade regime is found when custom duty is used as the trade liberalisation variable (both single entry in Model 2 and combined entry with trade variables in Models 7, 9 and 11) with the range of coefficients being 0.328 to 0.371. However, using tariff incidence as globalisation variable, no significant result is found. This means raising domestic protection by means of increasing custom duty rate by 1 per cent would increase employment by 0.328 per cent, other things held constant. On the other hand, when trade variables are used alone, significant negative coefficients are found in all cases (Model 4, 5 and 6) and the coefficient lies between -0.099 to -0.066. These results suggest that increase in trade activities result in a reduction in domestic employment which is consistent with the earlier findings that protection increases employment. However, combined entrance of any tariff variables with export-value added ratio and import-value added ratio as trade variables, causes the associated coefficients to become insignificant. When trade-value added ratio enters the model jointly with either of the tariff variables, the associated coefficients again become significant and negative.

Turning to the indirect impact of trade liberalisation, no evidence is found in Bangladesh context in favour of Rodrik's hypothesis of positive impact of globalisation on labour demand elasticities. The interaction terms between custom duty and wage are significantly negative while the interaction terms between trade variables and wage are significantly positive. In case of custom duty, negative interaction indicates that with higher real wage an increase in custom duty would yield a lower demand for labour. In other words, due to increase in custom duty the responsiveness of labour demand with respect to own price will increase. Similarly, positive interaction between trade and wage variable indicates an increase in trade activities which result in a higher demand for labour for higher wage (i.e., the responsiveness of labour demand with respect to own price will decrease as a result of an increase in trade activities).

The findings of this chapter in general, correlate the findings of the previous chapter. Increase in protection (by raising custom duties) is found to have positive employment implications (positive elasticity). Rodrik's hypothesis of positive impact of trade liberalisation on employment was not found to be the case for Bangladesh. Indeed, this has been the case for many of the developing countries as the liberalisation review in this chapter seems to indicate.

Annex A-6.1

Correlation Analysis among Selected Variables

Table A-6.1 displays Pearson's correlation coefficients among 12 variables over the period 1991-2000. Alongwith the pooled data, the analysis is also presented for export-oriented, non-export oriented, labour-intensive and non-labour intensive sectors.

As expected, gross output and value-added at factor cost show a strongly positive and significant correlation in all cases (within the range of 0.852 to 0.989). Both the gross output and value-added both have significant positive correlation with capital. However, in case of labour-intensive sector the relationship is very strong (0.985 for gross output while 0.978 for value-added). Similarly, number of production worker is significantly and positively correlated with non-production worker in all cases; however, for export-oriented sector the relationship is quite strong with a coefficient value of 0.905 followed by non-labour intensive sector (0.644) whereas in other cases, the relationship is relatively weak (0.369-0.480). The labour size (both production and non-production worker) is significantly and positively associated, although with varying degree of strength, with gross output, value-added at factor cost and capital stock which is quite logical from theoretical point of view.

In case of wage per employee, negative (in line with what the theory suggests) but insignificant coefficients are observed with both production and non-production workers for all models. However, for pooled data it reveals weak, negative but significant correlation with both tariff variables while relatively stronger positive relationship with import.

Quite naturally, both tariff variables (custom duty and tariff incidence) have no significant relationship with exports (measured by sectoral export orientation), however, have negative and significant relationship with import (measured by sectoral import penetration) for all sectors except labour-intensive ones. This indicates that labour-intensive goods are, in general, insensitive to changes in tariffs or protection level is still high to raise the import level. Some of the labour-intensive goods may be not major importables of Bangladesh currently.

It is interesting to note that in none of the cases, tariff variables show any kind of association with number of production workers; however, for non-labour intensive industries both tariff variables and for non-export sector only custom duty showed significant negative correlation with non-production workers.

On the other hand, in case of pooled, export and labour-intensive sectors, for all three trade variables, statistically significant and positive correlations are observed with production and non-production labour size, gross output, value-added at factor cost and capital stock whereas for non-export and non-labour intensive sectors, only import and total trade reveal significant positive correlation with these variables.

Annex Table A-6.1

Pearson's Correlation Matrices between Selected Variables for Different Sub-groups of Manufacturing Industries: 1991-2000

	Year	*PW*	*NPW*	*WPM*	*Gross*	*VAFC*	*Capital Stock*	*Custom Duty*	*Tariff Incidence*	*Export*	*Import*	*Trade*
A. All industries												
Year	1.000											
PW	0.046	1.000										
NPW	0.059**	0.405*	1.000									
WPM	-0.006	-0.025	-0.033	1.000								
Gross	0.075*	0.909*	0.382*	0.045	1.000							
VAFC	0.082*	0.882*	0.403*	0.077*	0.951*	1.000						
Capital stock	0.052***	0.326*	0.205*	0.107*	0.389*	0.473*	1.000					
Custom duty	-0.685*	0.026	-0.032	-0.050***	-0.020	-0.023	-0.102*	1.000				
Tariff incidence	-0.527*	0.025	0.010	-0.057***	0.006	0.002	-0.099*	0.852*	1.000			
Export	0.029	0.783*	0.212*	-0.023	0.746*	0.693*	0.163*	0.025	0.026	1.000		
Import	0.163*	0.249*	0.175*	0.144*	0.254*	0.247*	0.341*	-0.154*	-0.162*	0.087*	1.000	
Trade	0.068**	0.804*	0.244*	0.014	0.770*	0.718*	0.239*	-0.015	-0.016	0.969*	0.331*	1.000
B. Export sector												
Year	1.000											
PW	0.114	1.000										
NPW	0.079	0.905*	1.000									
WPM	-0.072	-0.055	-0.004	1.000								
Gross	0.143***	0.968*	0.864*	-0.059	1.000							
VAFC	0.137***	0.956*	0.907*	-0.033	0.975*	1.000						
Capital stock	0.167**	0.453*	0.664*	0.043	0.419*	0.519*	1.000					
Custom duty	-0.762*	0.011	0.047	0.119	-0.025	-0.008	-0.107	1.000				
Tariff incidence	-0.690*	0.048	0.063	0.083	0.009	0.030	-0.103	0.965*	1.000			
Export	0.112	0.934*	0.733*	-0.087	0.927*	0.879*	0.235*	0.007	0.057	1.000		
Import	0.133	0.243*	0.380*	0.015	0.261*	0.260*	0.494*	-0.176**	-0.190**	0.140***	1.000	
Trade	0.128	0.939*	0.772*	-0.079	0.935*	0.890*	0.317*	-0.024	0.019	0.983*	0.319*	1.000

contd...

...contd...

	Year	*PW*	*NPW*	*WPM*	*Gross*	*VAFC*	*Capital Stock*	*Custom Duty*	*Tariff Incidence*	*Export*	*Import*	*Trade*
C. Non-export sector												
Year	1.000											
PW	0.039	1.000										
NPW	0.056***	0.480*	1.000									
WPM	-0.002	-0.039	-0.034	1.000								
Gross	0.091*	0.386*	0.342*	0.171*	1.000							
VAFC	0.105*	0.385*	0.335*	0.215*	0.817*	1.000						
Capital stock	0.022	0.081*	0.083*	0.142*	0.389*	0.450*	1.000					
Custom duty	-0.677*	-0.004	-0.050***	-0.061***	-0.111*	-0.115*	-0.134*	1.000				
Tariff incidence	-0.508*	-0.009	0.000	-0.066**	-0.030	-0.050***	-0.122*	0.839*	1.000			
Export	-0.005	0.033	0.005	-0.012	0.002	0.008	0.018	0.013	0.003	1.000		
Import	0.183*	0.470*	0.131*	0.186*	0.341*	0.287*	0.237*	-0.165*	-0.172*	0.013	1.000	
Trade	0.064**	0.206*	0.053***	0.058***	0.129*	0.114*	0.105*	-0.050***	-0.062**	0.928*	0.386*	1.000
D. Labour-intensive sector												
Year	1.000											
PW	0.074	1.000										
NPW	0.083**	0.369*	1.000									
WPM	0.005	-0.016	-0.068	1.000								
Gross	0.071	0.972*	0.334*	0.004	1.000							
VAFC	0.077**	0.965*	0.375*	0.000	0.989*	1.000						
Capital stock	0.077**	0.974*	0.393*	0.004	0.985*	0.978*	1.000					
Custom duty	-0.693*	0.027	-0.019	-0.053	0.026	0.022	0.015	1.000				
Tariff incidence	-0.542*	0.036	0.062	-0.075	0.040	0.039	0.030	0.856*	1.000			
Export	0.063	0.961*	0.261*	-0.005	0.969*	0.955*	0.953*	0.028	0.039	1.000		
Import	0.135*	0.386*	0.246*	-0.026	0.266*	0.290*	0.289*	-0.068	-0.068	0.203*	1.000	
Trade	0.080***	0.979*	0.287*	-0.008	0.969*	0.959*	0.957*	0.016	0.027	0.990*	0.340*	1.000

contd...

...contd...

	Year	*PW*	*NPW*	*WPM*	*Gross*	*VAFC*	*Capital Stock*	*Custom Duty*	*Tariff Incidence*	*Export*	*Import*	*Trade*
E. Non-labour intensive sector												
Year	1.000											
PW	0.014	1.000										
NPW	0.041	0.644*	1.000									
WPM	-0.010	-0.010	0.017	1.000								
Gross	0.104*	0.598*	0.614*	0.113*	1.000							
VAFC	0.101*	0.669*	0.535*	0.157*	0.852*	1.000						
Capital stock	0.055	0.540*	0.380*	0.092*	0.525*	0.609*	1.000					
Custom duty	-0.681*	0.032	-0.077**	-0.040	-0.095*	-0.078**	-0.125*	1.000				
Tariff incidence	-0.519*	0.006	-0.085**	-0.045	-0.044	-0.039	-0.123*	0.850*	1.000			
Export	-0.008	0.048	0.023	-0.019	0.036	0.035	0.041	0.022	0.008	1.000		
Import	0.179*	0.287*	0.212*	0.163*	0.359*	0.271*	0.358*	-0.187*	-0.199*	0.020	1.000	
Trade	0.063***	0.156*	0.104*	0.046	0.174*	0.138*	0.178*	-0.054	-0.070**	0.920*	0.410*	1.000

Note: *,** & *** indicates correlation is significant at 1 per cent, 5 per cent & 10 per cent level respectively. The period of studies is from 1991 to 2000. PW, NPW, WPM, Gross, VAFC means production worker, non-production worker, wage per worker, gross output and value-added at factor cost respectively.

Source: Estimated from corrected real data from various CMI reports.

Annex Table A-6.2

Descriptions of Variables (Used in Various Regression Function)

Variables	*Production Function*	*Labour Demand Function*
lvafc (logarithm of value-added at factor cost)	Dependent	Independent
lpw (logarithm of production worker)	Independent	Dependent
lcs (logarithm of capital stock)	Independent	Independent
lcd (logarithm of custom's duty)	Independent	Independent
lti(logarithm value of tariff incidence)	Independent	Independent
lx_va(logarithm of sectoral export orientation)	Independent	Independent
lm_va(logarithm of sectoral import penetration)	Independent	Independent
lt_va(logarithm of sectoral trade orientation)	Independent	Independent
ltariff_wage 1 (interaction between custom duty and productive worker's wages)	-	Independent
ltariff_wage2 (interaction between tariff incidence and productive worker's wages)	-	Independent
ltrade_wage1 (interaction between export and wage)	-	Independent
ltrade_wage2 (interaction between import and wage)	-	Independent
ltrade_wage3 (interaction between trade with wage)	-	Independent
ltrend	Independent	Independent
ltrend_lcs (interaction between time trend and capital stock)	Independent	-
ltrend_pw (interaction between time trend and productive worker's wage)	-	-
l_gender (log of female participation rate)	-	-
exp_ind (export industries)	Independent	-
l_int (labour-intensive industries)	Independent	-

Annex A-6.2

Recovering the Production Function from the Labour Demand Function

The normalised cost function is $C=wl+K$ (A6.1)

From Shephard's Lemma and (6.6), we have

$$\frac{\delta C}{\delta w} = l = \exp(\alpha+\varepsilon)\, y^{\beta_y}\, g_1^{\beta_{g1}} g_2^{\beta_{g2}}\, t^{\beta_t} w^{[\beta_w + \beta_{wg1}\ln g1 + \beta_{wg2}\ln g2 + \beta_{wt}\ln t]}$$

So the (normalised) cost function must have the following restricted translog form:

$$C = \int_0^w \exp(\alpha+\varepsilon) y^{\beta_y} g_1^{\beta_{g1}} g_2^{\beta_{g2}} t^{\beta_t} z^{[\beta_w + \beta_{wg1}\ln g_1 + \beta_{wg2}\ln g_2 + \beta_{wt}\ln t]} dz$$
$$= \frac{\exp(\alpha+\varepsilon) y^{\beta_y} g_1^{\beta_{g1}} g_2^{\beta_{g2}} t^{\beta_t}}{[\beta_w + \beta_{wg1}\ln g_1 + \beta_{wg2}\ln g_2 + \beta_{wt}\ln t]+1} w^{[\beta_w + \beta_{wg1}\ln g_1 + \beta_{wg2}\ln g_2 + \beta_{wt}\ln t]+1} \quad \text{(A6.2)}$$

It is a restricted form in that the interaction term between output and wage and the squared wage do not enter the cost specification. Now singling out w yields

$$w = \left[\frac{l}{\exp(\alpha+\varepsilon) y^{\beta_y} g_1^{\beta_{g1}} g_2^{\beta_{g2}} t^{\beta_t}}\right]^{\frac{1}{[\beta_w + \beta_{wg1}\ln g_1 + \beta_{wg2}\ln g_2 + \beta_{wt}\ln t]}} \quad \text{(A6.3)}$$

Thus, substituting for c from (A6.2) and for w from (A6.3) into (A6.1) (after rearrangement) gives the desired production function as:

$$y = \left[\frac{l}{\exp(\alpha+\varepsilon) y^{\beta_y} g_1^{\beta_{g1}} g_2^{\beta_{g2}} t^{\beta_t}}\right]^{\frac{1}{\beta_y}} \left(-\frac{\in_{lw}}{1+\in_{lw}}\right)^{\frac{\in_{lw}}{\beta_y}} (k)^{-\frac{\in_{lw}}{\beta_y}} (l)^{\frac{1+\in_{lw}}{\beta_y}} \quad \text{(A6.4)}$$

where $\in_{lw} = \beta_w + \beta_{wg(1)} * \ln(g_{(1)}) + \beta_{wg(2)} * \ln(g_{(2)}) + \beta_{wt} * t$ as defined in (6.8). It is to be noted that the resulting production function is homothetic of degree $\frac{1}{\beta_y}$ and has a restricted translog form, with a variable technical efficiency given by

$$A = \left[\frac{l}{\exp(\alpha+\varepsilon) y^{\beta_y} g_1^{\beta_{g1}} g_2^{\beta_{g2}} t^{\beta_t}}\right]^{\frac{1}{\beta_y}} \left(-\frac{\in_{lw}}{1+\in_{lw}}\right)^{\frac{\in_{lw}}{\beta_y}} \quad \text{(A6.5)}$$

7 Trade Liberalisation and Sectoral Labour Demand in Bangladesh

A Computable General Equilibrium Analysis

In this chapter, an *ex ante* analysis of the impact of different trade liberalisations scenarios on the allocation of labour across different sectors is presented using a computable general equilibrium (CGE) model for Bangladesh.

The CGE analysis is the dominant methodology for the *ex ante* analysis of the economic consequences of comprehensive trade agreements whether multilateral or bilateral in nature (Francois and Shiells, 1994). This is the dominant methodology because no other approach offers the same flexibility for looking at prospective changes in trade policy while respecting the fundamental economy-wide consistency requirements such as balance of payments equilibrium and labour and capital market constraints that are so important in determining the consequences of comprehensive trade reforms.

7.1. Methodology

The general methodology has been to use a framework of analysis, which allows to examine the consequences of policy changes at sectoral and at macro level, and to estimate their poverty and distribution impacts at the representative households level. To examine the above issues a computable general equilibrium technique has been employed. A Social Accounting Matrix (SAM) prepared for the year 2005 serves as the consistent and comprehensive database for the above-mentioned exercises. In the present exercises, the 2005 SAM for Bangladesh has been treated as the base year data on the basis of which different simulation exercise have been carried out.

7.1.1. The Bangladesh Social Accounting Matrix 2005

As mentioned above, the model is numerically calibrated to a 2005 Social Accounting Matrix (SAM). The main sources of information for the SAM are: (a) 2000 SAM prepared by the International Food Policy Research Institute (IFPRI), Washington (Arndt *et al.*, 2002); (b) Bangladesh Economic Review (Ministry of Finance, 2006); (c) Export Receipts and Import Payments (Bangladesh Bank, 2006); and (d) National Income Estimates by Bangladesh Bureau of Statistics (May 2005).

Accounts

The 2005 SAM identifies the economic relations through four types of accounts: (i) production activity and commodity accounts for 26 sectors; (ii) nine factors of productions with 4 different types of labour and 5 types of capital; (iii) current account transactions between 4 main institutional agents; household members and unincorporated capital, corporation, government and the rest of the world; and (iv) two consolidated capital accounts to capture the flows of savings and investment by private and public institutions.

Activity and Commodity

The activity account is represented by 26 producing activities. A distinction is made between activity and commodity and the commodity account is denoted by 26 sectors.

Institutions Accounts

Current account transactions are captured between 4 institutional agents; households and unincorporated capital, corporate enterprise, government and the rest of the world. Household account includes 7 representative groups (e.g. 5 rural and 4 urban). Two consolidated capital accounts (domestic and rest of the world) distinguished by public and private sector origin are defined to capture the flows of savings and investment by institutions and the rest of the world respectively.

Household Members

The 2005 SAM distinguishes 7 household types, classified according to land holding size and occupation of the household's head in rural areas, and to level of education of the household's head in urban areas.

Labour Factor

The SAM includes 9 factors of production: land, ponds, non-agricultural capital, agricultural capital (further disaggregated into livestock and poultry) and 4 labour categories disaggregated by education levels and types of activity (agriculture and non-farm). The factor classification is based on the 21 factor classification used in the 1999-2000 IFPRI SAM for Bangladesh.

The disaggregation of factors, individuals, activities and institutions in the SAM and model is given in Table 7.1.

Table 7.1

Disaggregation and Description of Factors, Institutions and Households

Set	*Description of Elements*
Activities	
Agriculture (7)	Paddy, grains, other crops, livestock, poultry, fish and shrimp.
Industries (9)	Rice milling, grain milling, other food, mill clothing, woven RMG, knit RMG, textiles, petroleum products and other industries.
Services (10)	Urban construction, rural construction, public construction, utility, trade, transport, housing, education, health, public administration and private services.
Institutions	
Households (6)	Rural: Agricultural marginal, agricultural small, agricultural large, non-agricultural; Urban: Urban low educated, urban highly educated.
Others (2)	Government, corporation and rest of the world.
Factors of production	
Labour (4)	Agriculture labour unskilled, agriculture labour skilled, non-farm labour unskilled, non-farm labour skilled.
Capital (5)	Non-agriculture capital, land, ponds, agriculture capital poultry, agriculture capital cattle.

7.1.2. The CGE Model for Bangladesh Economy

Computable general equilibrium models capture detailed accounts of the circular flows of receipts and outlays in an economy. It satisfies general equilibrium conditions in market simultaneously. Such models are useful to analyse associations between various agents of the economy.

In line with most of CGE models, the model has been solved in comparative static mode and provides an instrument for controlled policy simulations and experiments. Solution of each simulation presents complete sets of socio-economic, meso and macro level indicators such as activity/commodity prices, household incomes and expenditures, factor demand and supplies, gross domestic products, exports and imports, and household poverty situation. The model is calibrated to the SAM to exactly reproduce the base year values.[1]

Activities

On the production side, it is assumed that in each sector there is a representative firm that generates value-added by combining labour and capital. A nested structure for production is adopted. Sectoral output is a Leontief function of value-added and total intermediate consumption. Value added is in turn represented by a CES function of capital and composite labour. The latter is also represented by a CES function of two labour categories: skilled labour and unskilled labour. Both labour categories are assumed to be fully mobile in the model. In the different production activities, we assume that a representative firm remunerates factors of production and pays dividends to households.

Household Members

They earn their income from production factors: labour, land and capital. They also receive dividends, intra-household transfers, government transfers and remittances. They pay direct income tax to the government. Household savings are a fixed proportion of total disposable income.

1. In calibration procedure, most of the model parameters are estimated endogenously keeping the various elasticity values fixed.

Household demand is derived from a C-D utility function. The model includes 4932 household members from the household survey.

Firms

There is one representative firm which earns capital income, pays dividends to households and foreigners and pays direct income taxes to the government.

Foreign Trade

It is assumed that foreign and domestic goods are imperfect substitutes. This geographical differentiation is introduced by the standard Armington assumption with a constant elasticity of substitution function (CES) between imports and domestic goods. On the supply side, producers make an optimal distribution of their production between exports and domestic sales according to a constant elasticity of transformation (CET) function. Furthermore, a finite elasticity export demand function is assumed. Even if it is assumed that the international terms of trade are given, the small country assumption for Bangladesh is rejected, and assumed that foreign demand for Senegalese exports is less than infinite. In order to increase their exports, local producers must decrease their free on board (FOB) prices.

Government

The government receives direct tax revenue from households and firms and indirect tax revenue on domestic and imported goods. Its expenditure is allocated between the consumption of goods and services (including public wages) and transfers. The model accounts for indirect or direct tax compensation in the case of a tariff cut.

System Constraints and Equilibrium Conditions

There are four constraints in the system. The real constraint refers to domestic commodity and factor market; the nominal constraint represents two macro balances: the current account balance of the rest of the world and the savings-investment balance.

Sectoral supply is a composite of imports and output sold in the domestic market. Composite demand, on the other hand, includes final demands (i.e., private and public consumption expenditure and investment) and intermediate input demand. Variations in the sectoral prices assure equilibrium between sectoral supply and demand.

In the case of factor market, we consider two variants of the model. In the 'full employment' model, it is assumed that total quantities of factors supply are fixed and factors are fully employed. This specification also implies full mobility of labour factors across producing activities and variations in their returns (e.g. wages) assures equilibrium in the factor market. On the other hand, the 'unemployment' specification of the model assumes that certain proportion of the labour force remain unemployed in the economy. In the 'unemployment' specification, no distinction is, however, made with respect to the percentage of labour factors unemployed for all labour categories.

The inflows (transfers to and from domestic institutions) are fixed but imports and exports are determined endogenously in the model. Foreign savings is fixed in this model and exchange rate acts as numeraire.

Finally, for the saving-investment equilibrium, the model treats the investment decision as given and hence, savings has to adjust to ensure the equality to the fixed value of investment. The basic approach is to allow the savings propensity of one of the domestic institution to vary.

7.2. Simulation Set-ups

In the simulation set-ups, two variations of the model are taken into consideration. Under both the 'full employment' and 'unemployment' models, three simulations are considered. The simulation set-ups are discussed below. It is important to note that, in CGE models there are concerns regarding the choice of the method to compensate the loss of revenue from a tariff reduction. Results of the simulation may vary considerably due to different compensation methods. Enhancing the production or income taxes in this regard can affect the relative prices, and thus can create further distortions in the economy. In the present exercise, the resulting loss in government revenue due to tariff elimination is compensated by a 'neutral tax adjustment'. This neutral tax is imposed on both the domestic goods and imported final goods and thus, it does not distort the relative prices in the import demand function.

7.2.1. Setting Up the Full Liberalisation Scenario

Under this scenario, tariffs on all imports are reduced to zero. The base values of all other parameters are retained.

7.2.2. Setting Up the Partial Liberalisation Scenario

Under this simulation, tariffs on all imports are reduced by 50 per cent. The base values of all other parameters are retained.

7.2.3. Setting Up the Rationalised Liberalisation Scenario

In the third simulation, tariff rates are rationalised in a way that conform to the tariff reforms' steps undertaken in the country. The base year tariff rates, as specified in the SAM 2005 of Bangladesh, are other crops (12.18 per cent), milled rice (9.42 per cent), other grains (27.98 per cent), food (40.18 per cent), other textile (3.53 per cent) and other industry (20.85 per cent). In this simulation, tariff rates on 'other crops' and 'other grains' are reduced by 50 per cent, tariff rate on 'milled rice' is reduced to zero, tariff rate on 'food' is reduced by 25 per cent, tariff rate on 'other textile' is left unaffected and finally, the tariff rate on 'other industry' is reduced down to 15 per cent. The base values of all other parameters are retained.

7.3. Simulation Outcomes

In this section, results of the simulations are presented. The major determinants of the simulation results are the values of trade elasticities, the share of imports and exports, the cost of inputs, and the general equilibrium effects of supply and demand. In all these cases, a decline in tariff rates leads to a decline in the production of the tariff protected sectors. The resources will then move from this sector to other expanding sectors. The changes in export and import prices influence the domestic and composite good prices, the value added price and determine factors' reallocation.

7.3.1. Macroeconomic Effects

The macroeconomic impacts are reported in Table 7.2.

Table 7.2

Macroeconomic Effects (% Change from the Base Year Value)

Variable	*"Full Employment" Model*			*"Unemployment" Model*		
	Simulation 1	*Simulation 2*	*Simulation 3*	*Simulation 1 (a)*	*Simulation 2 (a)*	*Simulation 3 (a)*
Real GDP	1.65	0.77	0.29	1.15	0.59	0.19
Agriculture	0.03	0.01	0.01	0.02	0.01	0.01
Manufacturing	5.01	2.12	1.09	3.65	1.61	0.89
Service	-1.45	-0.62	-0.31	-1.10	-0.47	-0.24
Traded	3.54	1.52	0.77	2.69	1.16	0.59
Non-traded	-1.58	-0.68	-0.31	-1.20	-0.52	-0.24
Consumption	1.35	0.65	0.34	1.03	0.39	0.26
Imports	14.87	6.75	3.07	11.30	5.80	2.33
Exports	27.46	12.42	5.64	18.87	9.44	4.29
Consumer price index	-0.12	-0.05	-0.03	-0.091	-0.038	-0.023
Wage rate of agri. unskilled labour	-0.068	-0.032	-0.019	-0.052	-0.024	-0.014
Wage rate of agri. skilled labour	-0.066	-0.032	-0.018	-0.053	-0.024	-0.013
Wage rate of non-agri. unskilled labour	-0.053	-0.025	-0.014	-0.044	-0.019	-0.011
Wage rate of non-agri. skilled labour	-0.055	-0.026	-0.015	-0.042	-0.020	-0.012

Note: 1. Real GDP is equal to the sum of consumption, investment, government consumption plus exports less imports in real terms for all sectors in the economy.

2. 2005 is the base year. Simulation outcomes are compared to base values.

Source: Authors' calculations, based on simulation results.

Full Employment Model

Under simulation 1, a full trade liberalisation will lead to an incremental GDP growth of 1.65 per cent. The aggregate real GDP declines are analysed further by three wide sectors such as agriculture, manufacturing and services. It appears that agriculture registers a very low growth, whereas manufacturing sector experience a high positive growth and the growth in manufacturing sector is much higher than that of real GDP. However, the service sector suffers from negative growth. The traded sector in the economy experiences a positive growth at the expense of the non-traded sector. There is a positive growth in real consumption. Exports experience a growth rate of 27.5 per cent, and also imports increases by around 15 per cent. This pattern is consistent with the structure of exports and imports in Bangladesh, where the major export items such as woven and knit RMG's output relies largely on imported raw materials. There is a fall in consumer price index by 12 per cent. Wage rate for all the labour categories declines, though the fall in wage rates is more pronounced in the agricultural sector compared to the non-agricultural sector. In simulation 2, under a partial liberalisation scenario when all tariffs are reduced by 50 per cent, the pattern of macroeconomic impacts is much similar to those under simulation 1. However, the impacts are less pronounced than those under simulation 1. Also, under simulation 3, the pattern of macroeconomic impacts is similar to those under simulation 1 and simulation 2. But the impacts under simulation 3 are the least among the three simulations under consideration.

Unemployment Model

When labour factors are not fully employed in the economy, then labour released from the shrinking sectors can't be fully absorbed in the expanding sectors. And as a result, the potentials of trade liberalisation can't be fully realised. In the context of the Bangladesh economy, where unemployment is a common phenomenon, it is more likely that the assumption of unemployment in the model will lead to different simulation outcomes compared to those under a 'full employment' model.

All three simulations are carried out under the 'unemployment' model and the macro results are reported in Table 7.2. It appears that though the patterns of the macro impacts are similar to those under the 'full employment' model, the presence of unemployment in the economy will likely to trim down the impacts. More specifically, the growths in imports and exports are less pronounced under this specification for all three simulations.

7.3.2. Impact on Sectoral Prices and Output

Full Employment Model

The impacts on sectoral prices and sectoral outputs are reported in Table 7.3 and Table 7.4 respectively. Under simulation 1, because of full elimination of tariff rates the import prices of the import-competing sectors decline significantly. It is anticipated that the sectors with higher tariff

Table 7.3

Percentage Changes in Prices from the Base-run ("Full Employment" Model)

	Simulation 1						Simulation 2						Simulation 3					
	PM	PD	PV	PX	PQ	PE_FOB	PM	PD	PV	PX	PQ	PE_FOB	PM	PD	PV	PX	PQ	PE_FOB
Paddy		-6.56	-6.27	-6.56	-6.56			-3.14	-2.95	-3.14	-3.14			-1.84	-1.83	-1.84	-1.84	
Grains		-6.28	-6.53	-6.28	-1.86			-3.02	-3.08	-3.02	-0.93			-1.74	-1.87	-1.74	-0.63	
Other crops	-10.85	-6.04	-6.07	-5.99	-4.33	-1.13	-5.43	-2.88	-2.86	-2.86	-2.11	-0.54	-5.43	-1.75	-1.79	-1.73	-1.53	-0.30
Livestock		-6.74	-7.04	-6.74	-4.71			-3.20	-3.30	-3.20	-2.26			-1.89	-1.96	-1.89	-1.43	
Poultry		-6.63	-7.13	-6.63	-6.63			-3.19	-3.37	-3.19	-3.19			-1.81	-1.91	-1.81	-1.81	
Shrimp		-10.86	-6.06	-7.52	-10.86	-1.61		-5.17	-2.88	-3.62	-5.17	-0.76		-2.81	-1.66	-1.99	-2.81	-0.43
Other fish		-6.28	-6.43	-6.22	-4.26	-0.92		-3.00	-3.05	-2.97	-2.08	-0.44		-1.69	-1.71	-1.67	-1.24	-0.25
Rice mill	-8.61	-6.34	-5.39	-6.34	-4.41		-4.31	-3.04	-2.54	-3.04	-2.17		-8.61	-1.77	-1.49	-1.77	-1.59	
Grain mill	-21.87	-3.39	-5.39	-3.39	-1.65		-10.93	-1.65	-2.54	-1.65	-0.87		-10.93	-1.00	-1.49	-1.00	-0.72	
Food	-28.67	-7.65	-5.33	-6.72	-10.84	-0.73	-14.33	-3.68	-2.56	-3.25	-5.14	-0.36	-7.17	-2.13	-1.50	-1.89	-2.77	-0.24
Mill cloth		-4.04	-5.35	-4.04	-4.04			-2.01	-2.57	-2.01	-2.01			-0.98	-1.50	-0.98	-0.98	
Woven RMG		-13.17	-5.31	-4.37	-11.19	-2.39		-6.50	-2.55	-2.16	-5.55	-1.13		-3.23	-1.49	-1.06	-2.76	-0.53
Knit RMG		-16.13	-5.31	-3.43	0.89	-3.35		-8.28	-2.55	-1.72	0.35	-1.68		-3.95	-1.49	-0.78	0.17	-0.76
Other textile	-3.41	-4.74	-5.37	-4.69	-2.10	-2.36	-1.71	-2.32	-2.57	-2.30	-1.12	-1.16		-1.36	-1.50	-1.35	-0.32	-0.60
Other industry	-17.25	-7.77	-5.32	-7.30	-10.34	-0.62	-8.63	-3.79	-2.56	-3.57	-5.12	-0.31	-4.14	-2.05	-1.49	-1.93	-2.56	-0.19
Urban construc.		-7.07	-5.30	-7.07	-7.07			-3.45	-2.55	-3.45	-3.45			-1.88	-1.49	-1.88	-1.88	
Rural construc.		-5.87	-5.28	-5.87	-5.87			-2.85	-2.54	-2.85	-2.85			-1.66	-1.49	-1.66	-1.66	
Public construc.		-7.76	-5.29	-7.76	-7.76			-3.79	-2.54	-3.79	-3.79			-2.01	-1.49	-2.01	-2.01	
Utility		-5.86	-5.33	-5.86	-5.86			-2.83	-2.56	-2.83	-2.83			-1.61	-1.50	-1.61	-1.61	
Trade		-5.80	-5.36	-5.80	-5.80			-2.79	-2.57	-2.79	-2.79			-1.60	-1.50	-1.60	-1.60	
Transport		-6.18	-5.33	-6.18	-6.18			-2.98	-2.55	-2.98	-2.98			-1.67	-1.49	-1.67	-1.67	
Housing		-5.48	-5.28	-5.48	-5.48			-2.64	-2.54	-2.64	-2.64			-1.53	-1.49	-1.53	-1.53	
Edu & health		-5.92	-5.40	-5.92	-5.92			-2.85	-2.59	-2.85	-2.85			-1.62	-1.51	-1.62	-1.62	
Pub admin.		-6.05	-5.42	-6.05	-6.05			-2.92	-2.60	-2.92	-2.92			-1.65	-1.51	-1.65	-1.65	
Pri. service		-5.80	-5.39	-5.80	-5.80			-2.79	-2.58	-2.79	-2.79			-1.60	-1.51	-1.60	-1.60	

Note: 1. PD = domestic goods price, PV = value-added price, PX = aggregate output price, PQ = price of composite goods, PE_FOB = FOB export price, PM = Domestic price of imports

2. 2005 is the base year.

Source: Authors' calculations, based on simulation results.

Table 7.4

Percentage Changes in Volumes from the Base-run ("Full Employment" Model)

	Simulation 1					*Simulation 2*					*Simulation 3*				
	M	*X*	*E*	*Q*	*D*	*M*	*X*	*E*	*Q*	*D*	*M*	*X*	*E*	*Q*	*D*
Paddy		-2.67		-0.58	-2.67		-1.15		-0.20	-1.15		-0.65		-0.20	-0.65
Grains	-10.06	1.09		-3.34	1.09	-4.75	0.65		-1.47	0.65	-2.71	0.42		-0.81	0.42
Other crops	12.37	2.34	12.05	2.80	2.22	6.02	1.12	5.52	1.36	1.07	7.49	0.39	3.04	0.77	0.36
Livestock	-12.77	-1.09		-1.14	-1.09	-6.06	-0.40		-0.42	-0.40	-3.59	-0.23		-0.24	-0.23
Poultry		-0.92		1.22	-0.92		-0.35		0.60	-0.35		-0.15		0.30	-0.15
Shrimp		5.26	17.67	0.63	-1.49		2.36	7.91	0.38	-0.57		1.50	4.43	0.42	-0.03
Other fish	-11.67	-0.62	9.71	-0.74	-0.74	-5.63	-0.25	4.49	-0.30	-0.30	-3.17	-0.12	2.50	-0.15	-0.15
Rice mill	3.24	-0.74		-0.59	-0.74	1.82	-0.29		-0.22	-0.29	11.97	-0.25		0.18	-0.25
Grain mill	35.76	-3.33		-2.80	-3.33	15.48	-1.45		-1.21	-1.45	17.46	-0.82		-0.56	-0.82
Food	44.96	-2.87	7.65	4.98	-4.10	18.52	-1.19	3.71	2.23	-1.75	8.08	-0.36	2.42	1.09	-0.68
Mill cloth		-3.36		-1.28	-3.36		-1.46		-0.52	-1.46		-0.95		-0.50	-0.95
Woven RMG	-9.32	23.53	26.92	8.77	8.95	-4.50	10.54	12.06	4.14	4.22	-2.47	4.76	5.48	1.74	1.78
Knit RMG	-6.96	40.48	40.63	-5.45	16.95	-3.31	18.35	18.42	-2.55	8.19	-1.85	7.88	7.91	-1.50	3.42
Other textile	21.17	23.45	27.35	22.40	23.37	9.70	10.63	12.31	10.20	10.60	3.29	5.17	6.20	4.33	5.15
Other industry	11.59	-3.51	6.42	2.74	-4.14	5.45	-1.61	3.11	1.37	-1.90	2.26	-0.62	1.88	0.59	-0.78
Urban construc.		1.42		3.60	1.42		0.76		1.73	0.76		0.31		0.76	0.31
Rural construc.		0.26		2.42	0.26		0.23		1.19	0.23		0.08		0.53	0.08
Public construc.		1.82		4.01	1.82		0.98		1.95	0.98		0.38		0.83	0.38
Utility		0.17		2.32	0.17		0.18		1.14	0.18		0.07		0.53	0.07
Trade		-0.98		1.16	-0.98		-0.33		0.62	-0.33		-0.25		0.20	-0.25
Transport		-0.54		1.60	-0.54		-0.16		0.80	-0.16		-0.12		0.33	-0.12
Housing		-1.53		0.59	-1.53		-0.64		0.31	-0.64		-0.31		0.14	-0.31
Edu & health		0.14		2.29	0.14		0.03		0.98	0.03		0.01		0.46	0.01
Pub admin.		1.50		3.69	1.50		0.61		1.57	0.61		0.26		0.71	0.26
Pri. service		-1.14		0.99	-1.14		-0.45		0.50	-0.45		-0.23		0.22	-0.23

Note: 1. M = imports, X = domestic Output, E = exports, Q = composite goods, D = domestic sales.
2. 2005 is the base year.

Source: Authors' calculations, based on simulation results.

rates are likely to experience higher fall in import prices. In fact, the sector with the highest tariff rate is the 'food' sector, and this sector experiences the fall in import prices by 28.7 per cent. Import price in the 'other industry' sector declines by 17.3 per cent. Because of the fall in import prices, the domestic sales prices as well as the composite prices in the import-competing sectors decline. Also, domestic sales prices and the composite prices in all other sectors decline. It is interesting to observe that though the output prices and the value-added prices in all sectors decline, the import-competing and the non-tradable sectors (both in agriculture and service sectors) experience larger fall than the export-oriented sectors. This suggests that this liberalisation scenario is likely to help the manufacturing and agricultural sectors which are export-oriented. The fall in the FOB export prices of the export-oriented sectors also indicate the possibility of expansion of these sectors. The fall in FOB export prices are the highest for the woven RMG and knit RMG sectors, which suggest that these two sectors are likely to experience significant expansion during the full liberalised regime.

It appears from Table 7.4 that due to the full elimination of tariffs on the import-competing sectors, there is a surge in imports into these sectors. For example, the 'food' sector experiences the rise in imports by as high as 45 per cent. Domestic production in most of the import-competing and non-tradable sectors declines. On the other hand, domestic production as well as exports from the export-oriented sectors increases significantly. Especially, the expansion of the 'woven RMG' and 'knit RMG' is noteworthy. Since these two sectors constitute bulk to the export earnings of the country, such high growth rates in these sectors helps the country achieve high growth in overall exports. The 'other textile' sector also expands because of increased demand of input from woven and knit RMG sectors, as well as the increase in exports from this sector.

The sectoral impacts on prices and volumes under simulations 2 and 3, are similar to those under simulation 1. However, the impacts are less pronounced than those under simulation 1. Under both simulations 2 and 3, the largest expanding sectors are 'woven RMG', 'knit RMG' and 'other textiles'.

Unemployment Model

Results of the simulation exercises under the 'unemployment' version for the impacts on sectoral prices and volumes are reported in Tables 7.5 and 7.6 respectively. Like the macro impacts, the patterns of sectoral impacts are also similar to those under the 'full employment' version of the model. However, the expansion of the export-oriented sectors, especially that of 'woven RMG' and 'knit RMG' sectors appears to be less pronounced. On the other hand, the import-competing sectors also contract by lesser magnitudes.

Overall, according to the CGE exercise, there appears to be a trade-off between the possibility of incremental labour absorption in export-oriented sectors and reduction of employment in import-competing sectors, as a consequence of further trade reforms in the context of Bangladesh. Thus, any further trade liberalisation in Bangladesh should be undertaken by carefully taking into cognisance the positive as well as adverse implications of such initiatives for the job market of the country.

7.3.3. Impact on Sectoral Labour Demand

Full Employment Model

As has been mentioned before, labour factor has been disaggregated into four categories: agricultural unskilled labour, agricultural skilled labour, non-agricultural skilled labour and non-agricultural unskilled labour. It is anticipated that due to trade liberalisation, labour in contracting sectors are likely to release labour, and because of the assumption of 'full employment', the released labour will be absorbed in the expanding sectors. We assume that there is a full mobility of agricultural labour factors across the production sub-sectors within the agricultural sector. Similarly, there is a full mobility of non-agricultural labour factors across the production sub-sectors within the non-agricultural sector.

It appears from Table 7.7 that as under simulation 1, among the agricultural sectors, as 'paddy', 'livestock' and 'poultry' sectors contract, demand for both skilled and unskilled agricultural labour decline in these sectors. On the other hand, demand for skilled and unskilled agricultural labour increases in all other agricultural sectors. The 'shrimp' and 'other

Table 7.5

Percentage Changes in Prices from the Base-run ("Unemployment" Model)

	Simulation 1 (a)						Simulation 2 (a)						Simulation 3 (a)					
	PM	PD	PV	PX	PQ	PE_FOB	PM	PD	PV	PX	PQ	PE_FOB	PM	PD	PV	PX	PQ	PE_FOB
Paddy		-5.90	-5.64	-5.90	-5.90			-2.83	-2.66	-2.83	-2.83			-1.66	-1.65	-1.66	-1.66	
Grains		-5.15	-5.35	-5.15	-1.53			-2.48	-2.53	-2.48	-0.76			-1.43	-1.53	-1.43	-0.52	
Other crops	-9.22	-5.44	-5.46	-5.39	-3.90	-1.02	-4.21	-2.59	-2.57	-2.57	-1.90	-0.49	-4.44	-1.58	-1.61	-1.56	-1.38	-0.27
Livestock		-5.53	-5.77	-5.53	-3.86			-2.62	-2.71	-2.62	-1.85			-1.55	-1.61	-1.55	-1.17	
Poultry		-5.97	-6.42	-5.97	-5.97			-2.87	-3.03	-2.87	-2.87			-1.63	-1.72	-1.63	-1.63	
Shrimp		-8.91	-4.97	-6.17	-8.91	-1.32		-4.24	-2.36	-2.97	-4.24	-0.62		-2.30	-1.36	-1.63	-2.30	-0.35
Other fish		-5.65	-5.79	-5.60	-3.83	-0.83		-2.70	-2.75	-2.67	-1.87	-0.40		-1.52	-1.54	-1.50	-1.12	-0.23
Rice mill	-7.32	-5.20	-4.42	-5.20	-3.62		-3.93	-2.49	-2.08	-2.49	-1.78		-7.53	-1.45	-1.22	-1.45	-1.30	
Grain mill	-18.59	-3.05	-4.85	-3.05	-1.49		-8.53	-1.49	-2.29	-1.49	-0.78		-8.75	-0.90	-1.34	-0.90	-0.65	
Food	-24.37	-6.27	-4.37	-5.51	-8.89	-0.60	-11.43	-3.02	-2.10	-2.67	-4.21	-0.30	-6.93	-1.75	-1.23	-1.55	-2.27	-0.20
Mill cloth		-3.64	-4.82	-3.64	-3.64			-1.81	-2.31	-1.81	-1.81			-0.88	-1.35	-0.88	-0.88	
Woven RMG		-10.80	-4.35	-3.58	-9.18	-1.96		-5.33	-2.09	-1.77	-4.55	-0.93		-2.65	-1.22	-0.87	-2.26	-0.43
Knit RMG		-14.52	-4.78	-3.09	0.80	-3.02		-7.45	-2.30	-1.55	0.32	-1.51		-3.56	-1.34	-0.70	0.15	-0.68
Other textile	-2.90	-3.89	-4.40	-3.85	-1.72	-1.94	-1.28	-1.90	-2.11	-1.89	-0.92	-0.95		-1.12	-1.23	-1.11	-0.26	-0.49
Other industry	-14.66	-6.99	-4.79	-6.57	-9.31	-0.56	-7.54	-3.41	-2.30	-3.21	-4.61	-0.28	-3.11	-1.85	-1.34	-1.74	-2.30	-0.17
Urban construc.		-5.80	-4.35	-5.80	-5.80			-2.83	-2.09	-2.83	-2.83			-1.54	-1.22	-1.54	-1.54	
Rural construc.		-5.28	-4.75	-5.28	-5.28			-2.57	-2.29	-2.57	-2.57			-1.49	-1.34	-1.49	-1.49	
Public construc.		-6.36	-4.34	-6.36	-6.36			-3.11	-2.08	-3.11	-3.11			-1.65	-1.22	-1.65	-1.65	
Utility		-5.27	-4.80	-5.27	-5.27			-2.55	-2.30	-2.55	-2.55			-1.45	-1.35	-1.45	-1.45	
Trade		-4.76	-4.40	-4.76	-4.76			-2.29	-2.11	-2.29	-2.29			-1.31	-1.23	-1.31	-1.31	
Transport		-5.56	-4.80	-5.56	-5.56			-2.68	-2.30	-2.68	-2.68			-1.50	-1.34	-1.50	-1.50	
Housing		-4.49	-4.33	-4.49	-4.49			-2.16	-2.08	-2.16	-2.16			-1.25	-1.22	-1.25	-1.25	
Edu & health		-5.33	-4.86	-5.33	-5.33			-2.57	-2.33	-2.57	-2.57			-1.46	-1.36	-1.46	-1.46	
Pub admin.		-4.96	-4.44	-4.96	-4.96			-2.39	-2.13	-2.39	-2.39			-1.35	-1.24	-1.35	-1.35	
Pri. service		-4.76	-4.42	-4.76	-4.76			-2.29	-2.12	-2.29	-2.29			-1.31	-1.24	-1.31	-1.31	

Note: 1. PD = domestic goods price, PV = value-added price, PX = aggregate output price, PQ = price of composite goods, PE_FOB = FOB export price. PM = Domestic price of imports

2. 2005 is the base year.

Source: Authors' calculations, based on simulation results.

Table 7.6

Percentage Changes in Volumes from the Base-run ("Unemployment" Model)

	Simulation 1 (a)					Simulation 2 (a)					Simulation 3 (a)				
	M	*X*	*E*	*Q*	*D*	*M*	*X*	*E*	*Q*	*D*	*M*	*X*	*E*	*Q*	*D*
Paddy		-2.32	0.00	-0.50	-2.32		-1.00	0.00	-0.17	-1.00		-0.57	0.00	-0.17	-0.57
Grains	-8.75	0.95	0.00	-2.91	0.95	-4.13	0.57	0.00	-1.28	0.57	-2.36	0.37	0.00	-0.70	0.37
Other crops	10.14	1.92	9.88	2.30	1.82	4.94	0.92	4.53	1.12	0.88	6.14	0.32	2.49	0.63	0.30
Livestock	-10.47	-0.89	0.00	-0.93	-0.89	-4.97	-0.33	0.00	-0.34	-0.33	-2.94	-0.19	0.00	-0.20	-0.19
Poultry		-0.80	0.00	1.06	-0.80		-0.30	0.00	0.52	-0.30		-0.13	0.00	0.26	-0.13
Shrimp		4.58	15.37	0.55	-1.30		2.05	6.88	0.33	-0.50		1.31	3.85	0.37	-0.03
Other fish	-9.57	-0.51	7.96	-0.61	-0.61	-4.62	-0.21	3.68	-0.25	-0.25	-2.60	-0.10	2.05	-0.12	-0.12
Rice mill	2.66	-0.61	0.00	-0.48	-0.61	1.49	-0.24	0.00	-0.18	-0.24	9.82	-0.21	0.00	0.15	-0.21
Grain mill	31.11	-2.90	0.00	-2.44	-2.90	13.47	-1.26	0.00	-1.05	-1.26	15.19	-0.71	0.00	-0.49	-0.71
Food	39.12	-2.50	6.66	4.33	-3.57	16.11	-1.04	3.23	1.94	-1.52	7.03	-0.31	2.11	0.95	-0.59
Mill cloth		-2.76	0.00	-1.05	-2.76		-1.20	0.00	-0.43	-1.20		-0.78	0.00	-0.41	-0.78
Woven RMG	-7.64	19.29	22.07	7.19	7.34	-3.69	8.64	9.89	3.39	3.46	-2.03	3.90	4.49	1.43	1.46
Knit RMG	-6.06	35.22	35.35	-4.74	14.75	-2.88	15.96	16.03	-2.22	7.13	-1.61	6.86	6.88	-1.31	2.98
Other textile	18.42	20.40	23.79	19.49	20.33	8.44	9.25	10.71	8.87	9.22	2.86	4.50	5.39	3.77	4.48
Other industry	9.50	-2.88	5.26	2.25	-3.39	4.47	-1.32	2.55	1.12	-1.56	1.85	-0.51	1.54	0.48	-0.64
Urban construc.		1.16	0.00	2.95	1.16		0.62	0.00	1.42	0.62		0.25	0.00	0.62	0.25
Rural construc.		0.23	0.00	2.11	0.23		0.20	0.00	1.04	0.20		0.07	0.00	0.46	0.07
Public construc.		1.58	0.00	3.49	1.58		0.85	0.00	1.70	0.85		0.33	0.00	0.72	0.33
Utility		0.14	0.00	1.90	0.14		0.15	0.00	0.93	0.15		0.06	0.00	0.43	0.06
Trade		-0.80	0.00	0.95	-0.80		-0.27	0.00	0.51	-0.27		-0.21	0.00	0.16	-0.21
Transport		-0.47	0.00	1.39	-0.47		-0.14	0.00	0.70	-0.14		-0.10	0.00	0.29	-0.10
Housing		-1.33	0.00	0.51	-1.33		-0.56	0.00	0.27	-0.56		-0.27	0.00	0.12	-0.27
Edu. & health		0.11	0.00	1.88	0.11		0.02	0.00	0.80	0.02		0.01	0.00	0.38	0.01
Pub. admin.		1.23	0.00	3.03	1.23		0.50	0.00	1.29	0.50		0.21	0.00	0.58	0.21
Pri. service		-0.93	0.00	0.81	-0.93		-0.37	0.00	0.41	-0.37		-0.19	0.00	0.18	-0.19

Note: 1. M = imports, X = domestic output, E = exports, Q = composite goods, D = domestic sales.
2. 2005 is the base year.

Source: Authors' calculations, based on simulation results.

Table 7.7

Percentage Changes in Labour Demand from the Base-run ("Full Employment" Model)

	Simulation 1				*Simulation 2*				*Simulation 3*			
	LAgUnsk	*LAgSk*	*LNAgUnsk*	*LNAgSk*	*LAgUnsk*	*LAgSk*	*LNAgUnsk*	*LNAgSk*	*LAgUnsk*	*LAgSk*	*LNAgUnsk*	*LNAgSk*
Paddy	-2.24	-2.55			-1.03	-1.15			-0.52	-0.65		
Grains	1.98	1.66			0.95	0.83			0.70	0.56		
Other crops	3.73	3.41			1.68	1.55			0.69	0.55		
Livestock	-0.95	-1.25			-0.37	-0.50			-0.16	-0.30		
Poultry	-0.60	-0.90			-0.31	-0.44			0.02	-0.11		
Shrimp	4.09	3.77			1.71	1.59			1.40	1.27		
Other fish	0.37	0.06			0.12	0.00			0.27	0.13		
Rice mill			-2.38	-2.11			-1.08	-0.94			-0.68	-0.62
Grain mill			-0.41	-0.14			-0.12	0.02			-0.10	-0.03
Food			-5.06	-4.80			-2.27	-2.14			-0.98	-0.91
Mill cloth			-1.44	-1.17			-0.66	-0.52			-0.20	-0.13
Woven RMG			25.35	25.69			11.17	11.33			5.44	5.51
Knit RMG			44.64	45.04			19.83	20.00			9.03	9.11
Other textile			24.62	24.96			11.03	11.18			5.38	5.46
Other industry			-6.56	-6.31			-3.17	-3.03			-1.31	-1.24
Urban construc.			-1.43	-1.16			-0.65	-0.52			-0.30	-0.23
Rural construc.			-0.65	-0.38			-0.26	-0.12			-0.19	-0.12
Public construc.			-2.13	-1.87			-0.97	-0.83			-0.44	-0.37
Utility			-0.73	-0.46			-0.28	-0.14			-0.12	-0.05
Trade			-1.78	-1.51			-0.72	-0.58			-0.43	-0.36
Transport			-1.93	-1.66			-0.85	-0.71			-0.41	-0.35
Housing												
Edu. & health			-0.85	-0.58			-0.47	-0.33			-0.21	-0.14
Pub. admin.			0.28	0.56				0.14				0.07
Pri. service			-1.93	-1.67			-0.85	-0.71			-0.41	-0.34

Note: 1. LAgUnsk = agricultural unskilled labour, LAgSK = agricultural skilled labour, LNAgUnsk = non-agricultural unskilled labour, LNAgSk = non-agricultural skilled labour.

2. 2005 is the base year.

Source: Authors' calculations, based on simulation results.

crops' sectors experience the largest increase in demand for both skilled and unskilled agricultural labour. In the case of manufacturing sector, as a result of the contraction of the sectors like 'rice milling', 'grain milling', 'food', 'mill cloth' and 'other industries', both skilled and unskilled non-agricultural labour from these sectors are released. The largest fall in growth in labour demand is experienced by the 'other industries' sector followed by the 'food' sector. On the other hand, three manufacturing sectors, i.e., 'woven RMG', 'knit RMG' and 'other textiles', experience high rise in demand for skilled and unskilled labour. The growth in labour demand in the 'knit RMG' sector is the highest. Finally, all the services sectors experience fall in demand for both skilled and unskilled labour.

The pattern of the results for simulation 2 and 3 are similar to those under simulation 1. However, as discussed before, the magnitudes of the impacts are smaller in these two simulations.

Unemployment Model

The simulation results for the changes in sectoral labour demand under the 'unemployment' version of the model are reported in Table 7.8. As the export-oriented sectors expand less prominently than those under the 'full employment' model, the growth in demand for labour also appears to be less pronounced.

7.4. Conclusion

This chapter investigated the impacts of different trade liberalisation scenarios on the macroeconomy as well as on sectoral prices, outputs, trade and labour demand. It appears that any further trade liberalisation in Bangladesh, in general, will favour the expansion of the export-oriented sectors at the cost of import-competing sectors. Such results hold under both the 'full employment' and 'unemployment' specifications of the CGE model.

In terms of policy implications, such results should not be interpreted to indicate that there is no need for further trade liberalisation in Bangladesh. There is certainly a need for rationalisation of the tariff regime which may require further liberalisation in some sectors. Some critically important sectors in the economy, at the same time, also need to be given

Table 7.8

Percentage Changes in Labour Demand from the Base-run ("Unemployment" Model)

	Simulation 1				*Simulation 2*				*Simulation 3*			
	LAgUnsk	*LAgSk*	*LNAgUnsk*	*LNAgSk*	*LAgUnsk*	*LAgSk*	*LNAgUnsk*	*LNAgSk*	*LAgUnsk*	*LAgSk*	*LNAgUnsk*	*LNAgSk*
Paddy	-1.84	-2.09			-0.84	-0.94			-0.43	-0.53		
Grains	1.72	1.44			0.83	0.72			0.61	0.49		
Other crops	3.39	3.10			1.53	1.41			0.63	0.50		
Livestock	-0.78	-1.03			-0.30	-0.41			-0.13	-0.25		
Poultry	-0.52	-0.78			-0.27	-0.38			0.02	-0.10		
Shrimp	3.72	3.43			1.56	1.45			1.27	1.16		
Other fish	0.34	0.05			0.11	0.00			0.25	0.12		
Rice mill			-1.95	-1.73			-0.89	-0.77			-0.56	-0.51
Grain mill			-0.36	-0.12			-0.10	0.02			-0.09	-0.03
Food			-4.60	-4.37			-2.07	-1.95			-0.89	-0.83
Mill cloth			-1.18	-0.96			-0.54	-0.43			-0.16	-0.11
Woven RMG			22.05	22.35			9.72	9.86			4.73	4.79
Knit RMG			40.62	40.99			18.05	18.20			8.22	8.29
Other textile			22.40	22.71			10.04	10.17			4.90	4.97
Other industry			-5.38	-5.17			-2.60	-2.48			-1.07	-1.02
Urban construc.			-1.24	-1.01			-0.57	-0.45			-0.26	-0.20
Rural construc.			-0.59	-0.35			-0.24	-0.11			-0.17	-0.11
Public construc.			-1.75	-1.53			-0.80	-0.68			-0.36	-0.30
Utility			-0.64	-0.40			-0.24	-0.12			-0.10	-0.04
Trade			-1.62	-1.37			-0.66	-0.53			-0.39	-0.33
Transport			-1.76	-1.51			-0.77	-0.65			-0.37	-0.32
Housing			0.00	0.00			0.00	0.00			0.00	0.00
Edu. & health			-0.74	-0.50			-0.41	-0.29			-0.18	-0.12
Pub. admin.			0.25	0.51			0.00	0.13			0.00	0.06
Pri. service			-1.76	-1.52			-0.77	-0.65			-0.37	-0.31

Note: 1. LAgUnsk = agricultural unskilled labour, LAgSK = agricultural skilled labour, LNAgUnsk = non-agricultural unskilled labour, LNAgSk = non-agricultural skilled labour.

2. 2005 is the base year.

Source: Authors' calculations, based on simulation results.

protection in consideration of various factors which have been mentioned in this study. The CGE modelling exercises suggest that a drastic tariff liberalisation, whether full or partial, would lead to significant employment loss in the import competing sectors which may not be reabsorbed fully in the export-oriented sectors (as in the case of 'unemployment' version of the model). This was likely to result in rise in unemployment in the economy. Consequently, a more pragmatic policy option would be not to go for any drastic tariff cut in the near term, but to undertake a gradual approach which should be an optimal mix of both liberalisation and protection. The arguments postulated in the new trade theories are very much in line with this later approach. Besides, cross-country experiences with the ongoing global financial crisis also testify to the need for a non-dogmatic, pragmatic approach with respect to the issue of further trade liberalisation in the context of developing countries such as Bangladesh.

8 Summary of Results and Policy Implications

Introduction

Implications for employment continue to remain a major concern in the context of trade liberalisation, particularly in developing countries with high un/underemployment. At one level, productive employment is a major route out of poverty and hence, in a country such as Bangladesh, it should be an issue of critical importance. There is a vast body of literature which shows that trade is good for economic growth which in turn should be good for poverty reduction. However, it is to be noted here that there are a number of assumptions behind this line of causation which does not hold when confronted with empirical evidence. First, while high rate of economic growth is a necessary condition for poverty reduction, experience has shown that it is not sufficient. The pattern of growth and in that context, the employment outcome of growth, is an important element in determining the poverty reducing outcome of growth. Second, while trade may be good for growth, it does not necessarily imply that trade liberalisation *per se* is good for growth. As a review of literature indicates, indeed, trade liberalisation, if pursued without due consideration to the strategic interests of a given economy, can have a negative impact on growth itself. Third, the impact of trade liberalisation on employment may not always be positive. And if that turns out to be the case in a country such as Bangladesh, then one will need to closely examine the related issues with a view to draw the relevant policy conclusions.

The present study has examined the changes in the employment scenario of Bangladesh following the pursuance of the trade liberalisation strategy and the possible effects of further trade liberalisation. The study has first reviewed the tariff reforms pursued in Bangladesh and traced the dynamics of changes in the relevant correlates since the early 1980s. Based on an examination of the speed of change in the major trade liberalisation related correlates and indicators, four phases of reform process in Bangladesh were identified with a view to subsequently establishing the interface and impact of the trade liberalisation on the country's labour market dynamics. Thus, the study has reviewed the change in the structure of employment in formal and informal sectors of Bangladesh during the pre and post-liberalisation period. A number of tools and techniques were used to establish the linkage between the relevant indicators and estimate the impact of trade liberalisation on employment. Firstly, data related correction was undertaken with regard to the data provided by the various LFSs to adjust for differences in definitional and age criteria with regard to labour force. Secondly, consistency and validity of the Bangladesh CMI data, at disaggregated level, was checked and necessary corrections were made in this context. Given the poor quality of labour force data in Bangladesh, and the attendant lack of comparability, this consistency and validity test was of utmost importance from the perspective of undertaking an exercise of this nature. Subsequent analyses were carried out based on the corrected datasets. Thirdly, sectoral trade and tariff data were estimated, and these variables were used to measure the extent of trade liberalisation in Bangladesh. Fourthly, the trade liberalisation correlates were juxtaposed with aggregate employment dynamics to examine the nature of impact of the former on the latter. Fifthly, econometric analyses (estimation of production function and labour demand function) were carried out using the 4-digit disaggregated industry-level data from successive CMI reports to provide more insightful results compared to those obtained by using the more aggregate level data. Finally, a CGE modelling exercise was carried out and a number of *ex ante* trade liberalisation policy scenarios were simulated with a view to exploring the likely impact of those scenarios on sectoral labour demand in the context of the Bangladesh economy.

Key Findings of the Study

The major findings of the study are summarised below:

- There has been significant liberalisation of trade in Bangladesh over the recent past period. Trade policies pursued in the 1990s compares quite starkly with those pursued in the 1970s and

1980s. Simple average MFN applied tariff rate, import weighted average tariff rate, implicit nominal tariff rate and quantitative restrictions declined significantly during the 1990s. For example, simple MFN applied average tariff and import weighted average tariff rates came down to 19.50 and 13.80 per cent respectively in 2000 from 88.60 and 42.10 per cent in 1991. Also, in terms of trade-GDP ratio there has been remarkable rise in the share over the last two decades; the ratio had crossed 40 per cent in 2006 from 17 per cent in 1991. Based on the pace of liberalisation, a number of distinctive phases had been identified in the present study. Indeed, the fastest pace of trade liberalisation in Bangladesh has taken place during the second phase (1991-1995).

- Overall, and in general, tariff liberalisation in Bangladesh has followed the classical sequence of higher protection for final goods and relatively lower protection for capital goods and intermediates. However, this was more prominent when nominal rates of protection were considered rather than in case of import weighted rates.
- Observing the sectoral and overall employment growth, it becomes evident that during the second phase of trade liberalisation, employment, in general, has suffered. In particular, agriculture and manufacturing sector (mainly the industrial sector) have experienced average negative growth in employment. On the other hand, recovering from the negative per annum growth in the first phase, service sector has experienced positive growth during the second and third phases.
- Throughout the period of analysis, manufacturing employment has remained stagnant while service sector posted a rising trend with a big jump in 1995-96; agriculture has experienced a decrease in 1995-96 following which employment in this sector has again increased. The share of agriculture in GDP has been declining over the years; however, the sector's contribution continues to be significant, contributing almost half of total employment in Bangladesh. Contribution of service sector to employment has experienced a decline in the late 80s but

increased beginning from the mid-90s. However, the share of manufacturing employment in total labour force shows an erratic pattern during the first and second phases. From the beginning of the third phase, manufacturing employment indicates an increasing trend. It is to be pointed out here that share of manufacturing sector in total employment is still very low and has not shown any increasing trend indicating the failure of manufacturing sector to absorb the growing labour force (in a Lewisian sense).

- Till 1999-2000, an increase in self-employed category is observed at the cost of reduction of employment in the unpaid family worker category. However, after 2000, there has been a drift back towards unpaid family workers. The share of other categories remained more or less unchanged for all the three phases.
- As revealed by the LFS data, there was an increase in total employment over the past years. On the other hand, the labour force was growing at a faster rate compared to the rate of creation of new job opportunities, resulting in a rise in unemployment rate. From the second phase of trade liberalisation, unemployment rate has been on the rise more rapidly for all cross-sections of the population. It is interesting to note that while open unemployment has risen, the degree of underemployment (by time measure) has fallen. The underemployment rate is lower in urban areas as compared to the rural areas.
- The overall female participation rate (FPR) was about 29 per cent in Bangladesh in 2005-06. No conclusive evidence could be discerned with regard to the impact of trade liberalisation on female participation in the labour force. Some sectors have observed a rise in FPR (e.g. RMG) while some others have experienced a fall in FPR (e.g. small and cottage industries).
- According to the PRSP, the projected total employment for FY 2006 was 51.69 million against a base of 44.03 million in FY 2003 (from LFS 2003). That would imply a compound annual growth rate (CAGR) of 5.49 per cent during this period. However, analysis of LFS data (2003 and 2006) reveals that the CAGR of

actual overall employment from FY 2003 to FY 2006 was 2.28 per cent, with the rural rate being 2.42 per cent and the urban rate being 1.84 per cent. The PRSP document has also set the target of total employment to be 58.08 million in FY 2008 against the base of FY 2003 implying a CAGR of 5.70 per cent between FY 2003 and FY 2008. Given the present employment scenario as reported in LFS of FY 2006, to achieve this projected employment size in FY 2008, overall employment has to be generated at the rate of 10.69 per cent, with rural being by 8.59 per cent and urban by 17.16 per cent annually over the next two years (if the actual employment figure for 2006 is taken into account). Given the current macroeconomic scenario and the recent employment and growth relationship experience, this will indeed be a Herculean task. Considering the recent trends of urbanisation in the country and slow pace of employment creation in urban areas, achieving the PRSP employment targets will be an issue of critical importance from the perspective of policy-making.

- As estimated from the LFS data, labour productivity for overall and manufacturing sector shows significant and strong positive correlations with corresponding real wages. However, in case of agriculture, no significant association between productivity and wages could be found. Lack of wage data restricts an examination of this relationship for the service sector. Considering the dominance of services in the overall economy, one could justifiably discern a positive alliance between wage and productivity in case of service sector as well.
- Although evidence is found favouring a linear association between real wage and productivity for the overall economy, as also for the manufacturing sector and service sectors and this can be reasonably explained, the growth in real wages lagged behind that of productivity. It is clearly visible from the study that these two indices diverge over time for both overall and manufacturing sectors; the gap between these is highest in case of the manufacturing sector. These findings lead to the following plausible hypotheses which are, of course, subject to empirical verification. The sectoral real wage elasticities with respect to

corresponding labour productivity would perhaps vary from sector to sector. In case of agriculture, the relative movements in real wage and productivity would suggest inelastic behaviour of agricultural real wages with respect to productivity rise. If productivity in agricultural sector increases and if wage does not respond positively to the increase in productivity, then it would imply that workers are not receiving enough gains from productivity rise. If the reverse happens, i.e., if wage remains sticky or increases when the productivity falls, it implies that the workers are getting protection against fall in productivity. In case of manufacturing sector the elasticity, although positive, would seem to be low compared to that of the overall economy. If wage remains sticky during the period of rising productivity in manufacturing sector, this would indicate that workers are not getting a share in the rising productivity through rising real wages. If the reverse happens in the manufacturing sector, i.e., if wage remains sticky or increases during the period of decreasing productivity, this may lead to a trade-off between real wage determination and fall in competitiveness.

- Data suggests that because of tariff liberalisation a number of import substituting industries, for example, sugar, yarn, edible oil, paper, pharmaceuticals, tobacco cigarette, cloth mill, chemical fertiliser, iron and steel, metal products, machinery and transport equipment industries have faced a decline in respective ERPs over the last two decades. However, not all import substituting industries have performed badly during this period. Sectors with negative growth were sugar, edible oil, paper, clothing mill and transport equipment industries whilst industries such as yarn, pharmaceuticals, cement, tobacco cigarette, chemical fertiliser, iron and steel and metal products experienced positive growth during that period.
- It was found from CMI data that export-oriented sectors were more protected than non-export oriented sectors and labour intensive sectors were more protected than non-labour intensive sector. It was also found that growth of production workers in

labour intensive sectors was higher than that of non-labour intensive sectors. For export-oriented sectors, the growth rates tend to be higher, with exceptions in particular years.

- It is important to note that, one of the major difficulties to assess the impact of trade liberalisation is to separate its impact from other policies and events or global market fluctuations whose impacts are likely to influence the impact of shocks originated from trade liberalisation. There are also a number of other policies and issues which are not directly treated as trade reform policies, rather they should be treated as accompanying policies to trade reform ones, that could have significant employment and wage implications of the country. Some of these effects could act in the same direction with trade liberalisation while others could act on opposite direction.
- Growth in employment of productive workers is observed for both export and non-export sectors, though the growth rate has been relatively higher in case of the export-oriented sectors. In general, during the liberalisation period, labour-intensive sectors have expanded more rapidly than the non-labour intensive sectors.
- Production function of the manufacturing industries in Bangladesh was estimated by using panel econometric techniques. In this connection, impact of trade liberalisation variables on the growth of output in the manufacturing sector was investigated. It appears that three trade liberalisation variables considered in this study, i.e., custom duty, tariff incidence and sectoral export orientation, turn out to be statistically insignificant in the regression model. On the other hand, two trade liberalisation variables i.e., sectoral import penetration and sectoral trade orientation have significant and robust negative impacts on output growth. These results would suggest that during the 1990s, increased import penetration and increased trade orientation (primarily through increased import penetration) did not have a positive impact on the output growth in the manufacturing sector of Bangladesh.

- Panel econometric estimation results for the labour demand function reveals some interesting results. Significant positive direct employment impact of more restricted trade regime is found when custom duty is used as the trade liberalisation variable; however, using tariff incidence as globalisation variable, no significant result is found. On the other hand, when trade variables are used alone, significant negative coefficients are found on labour demand in all cases. These results tend to suggest that increase in trade activities in Bangladesh have resulted in a reduction in domestic employment. These results also corroborate findings of a number of cross-country studies.
- Results of the different trade liberalisation simulation exercises using the CGE model for Bangladesh economy are also consistent with those found in the econometric estimation. The CGE modelling exercise shows that in terms of employment generation, there will be many sectors that will lose out as a result of trade liberalisation; some sectors will win as well. It appears that trade liberalisation would lead to expansion of the export-oriented manufacturing industries and as a result, these sectors would experience increase in employment under a more liberalised trade regime. However, in this context the performance of the non-export oriented sectors, especially those in the agriculture and service sectors, turn out to be unsatisfactory. The simulation exercise reveals that if further liberalisation is undertaken, the results are expected to be as those articulated above.

Based on the aforementioned analyses, a number of policy implications could be suggested:

Analysis suggests that, in general, there has been a rather weak relationship between trade liberalisation and growth in the manufacturing industries in Bangladesh. Trade liberalisation, whilst perhaps a necessary condition for industrial growth, did not prove to be a sufficient condition in the context of Bangladesh. In addition, other factors such as state of business support services, adequate supply of infrastructure, adequate supply of skilled labour, easy access to credit on reasonable terms, a conducive business environment, good governance and global business environment including market access situation are important in this context.

It is also seen that not all sectors are equally gainers or losers from opening up of the economy. Transferring of part of the gains from winners to the losers through redeployment of workers in expanding sectors from the contracting sectors is another important issue that needs to be considered from the policy maker's perspectives. As our empirical examination indicates, as far as trade liberalisation is concerned, such reform measures have not been able to increase labour demand for the manufacturing industries as a whole. Though it is true that the export oriented sectors and to some extent some non-export oriented sectors, have been able to generate employment opportunities during the liberalised trade regime, the fall in labour demand in other import-competing sectors has been large enough to result in a net negative effect. It is, therefore, important for the policy makers to design future trade liberalisation programme in such a way that it takes into account the need for employment generation in a large number of non-export oriented sector in the economy.

Because there are winners and losers as a consequence of trade reforms, policies will need to be designed for the losers as well. The findings of the study reinforces the argument for supportive actions favouring workers in the form of assistance to displaced workers and funds, skill upgradation, training for imparting new skills and various safety net programmes.

It is critically important to take into account the possible adverse social consequences of trade liberalisation and adopt measures to minimise, or if possible, to avoid such adverse consequences. The first such issue is the impact on total employment in an economy and the "churning" that takes place in the labour market of a country as a result of trade liberalisation. While there may be an adverse impact of trade liberalisation on employment as a whole, there may be differential effects on different segments of the labour market. A rise in the overall rate of unemployment is an example of the former type of effect. On the other hand, even when the overall unemployment situation does not deteriorate, some groups may lose out due to a decline of some industries while others benefit from new jobs that are created. In Bangladesh, we have examples of both (*viz.*, the decline of traditional industries like jute and cotton textiles, matches, etc., and the rise of new industries like garments). In that kind of a situation, one cannot simply forget those who are losing out just because in an overall sense, there has not been a drop in total employment. Appropriate measures would be needed to provide social protection and facilitate alternative sources of livelihood for those who might lose out in the process of adjustment.

Another social dimension is the type of skill and education that is required in the production structure that emerges after trade liberalisation. Demand for higher education and skill levels and new types of skills should be met by appropriate measures to impart such education and skills to the members of the labour force.

Employment and labour market implications of trade liberalisation may also have a gender dimension. While some newly emerging industries (e.g., export-oriented industries like garments, electronics, toys, etc.) may create demand for more women workers, decline in cottage and small scale industries, especially in rural areas (e.g., handlooms, food processing, etc.) may have an adverse effect on women's employment. Moreover, the quality of new employment that is generated for women is also an issue that needs to be addressed.

The above are only some of the possible social dimensions of trade liberalisation that a country like Bangladesh should address. A good way to address these would be to put in place appropriate social safety nets and introduce active labour market policies that would combine unemployment benefits, training and re-training, and assistance in job search. Creation of a "social fund" could be a good vehicle for implementing such measures.

Indeed, some initiatives had been taken in Bangladesh to address the adverse employment implications of trade liberalisation over the past few years through training of retrenched workers and skill upgradation. However, instead of being proactive initiatives a large part of such initiatives were more of a reactive nature. There is no clear and long term strategy as to how training needs and people needing training are to be identified, how redeployment of the trained people is to be made and how safety nets during the search period is to be organised. The rural sector is the major source of the supply of surplus labour and the urban sector is seen as the creator of demand for labour and employment opportunities. Addressing the best ways to bridge these labour demand and labour supply scenarios through appropriate incentives and initiatives continues to be a major challenge for Bangladesh. However, this social dimension and technical aspects of labour force absorption are often not given adequate importance in the employment policies.

It is important here to mention that, although liberalisation should encompass many factors affecting trade and business practices, in Bangladesh an overwhelming attention has been given to trade related instruments only. In fact, policy makers are so inclined to measures related to tariffs and QRs that most of the time reform measures are used interchangeably with trade liberalisation measures. Reforms of the institutions have largely been overlooked in this context. Enforcing such trade reforms as tariff cuts and elimination of QRs, are relatively easy. However, significant growth and employment enhancing effects perhaps require reforms in other difficult areas. In this regard, there are suggestions that institutional reforms should be considered as the key to Bangladesh's growth and employment supporting strategies. It is important that trade policy reform is considered as institutional reform, as has been argued strongly by Rodrik (2002). This is not to argue against any further trade liberalisation, but to emphasise the point that such measures should be seen as part of a comprehensive package that has employment considerations as an important element.

It is also important to recognise the difference between trade policy and policy of trade liberalisation. The distinction is important for protecting some policy space that is required to promote development priorities. An overall pro-liberalisation policy does not exclude the possibility of policy support for certain specific sectors. The strategic trade policy school has strongly argued in favour of such a nuanced approach. Across the board drastic tariff reduction may not only have revenue implication for the government, but this should also be seen in conjunction with the need to also provide strategic support to domestic industries with high growth, employment and poverty alleviation potentials. By adopting a proactive and analytical policy regime, effective support to the growth of small and informal sector activities with significant employment and poverty alleviation effects can be provided; however, to a large extent such opportunities appear to have been missed in Bangladesh in the absence of a clear-cut trade policy philosophy and lack of manoeuvrability and flexibility.

The analyses presented in the study indicate that further deepening of the trade reforms must be considered by weighing-in the possible adverse consequences for the labour market. The study shows that if employment is to be protected, the policy makers will need to harness the competitive strength of domestic market-oriented sectors so that these are able to compete with imported goods. Whilst other factors such as policy support, incentives, infrastructure are critical factors of competitiveness, tariffs, which in some cases, in spite of reforms, tends to be high in Bangladesh, still continue to play important role in providing protection to industries. As such, further across the board liberalisation may not be the best policy option for Bangladesh, particularly, given the adverse impact this could have on labour market. Policies that make production for exports more profitable than import-competing sector, but have adverse impact on overall employment situation in the country, should be pursued with caution. What transpires from the study, that any further trade reform should be pursued with caution and after assessing the overall impact on employment and social welfare. The analyses indicate the need for reorienting the policies favouring import-substituting industries, without, however, reverting back to the polices of the 1970s and 1980s. What is called for, is to craft further trade reform by taking cognisance of the implications and current experience. The global financial crisis should serve as a wake up call in this context.

The ongoing financial crisis provides a sobering opportunity to revisit the case of trade liberalisation in the context of developing countries such as Bangladesh. Countries which had strong domestic sectors, were relatively less exposed to global markets and had relatively higher share of domestic consumption in the GDP, were those which were relatively less affected by the crisis. As in most crises, it is the poor who has been hit the hardest and the earliest. In many countries, dependent on export markets, such as China, the impact on labour market has been devastating. The need to protect domestic industries and sectors through selective trade policies has once again been reinforced by the crisis. Whilst the crisis should not be seen as vindication of reverting back to the old dirigistic, inward-looking policies underwritten by high tariff wall, the experience of recent times also call for a careful examination of the policy of trade reforms and the possible pitfalls involved. This would call for careful sequencing of further trade reforms in developing countries such as Bangladesh.

References

Agenor, P. and J. Aizenman (1996). "Trade liberalisation and unemployment", *Journal of International Trade and Economic Development* 5(3): 265-286.

Ahmed, N. (2001). *Trade Liberalisation in Bangladesh: An Investigation into Trends.* Dhaka: The University Press Ltd.

Ahmed, S. and Z. Sattar (2003). "Trade liberalisation, growth and poverty reduction: The case of Bangladesh", *mimeo*. World Bank, Dhaka.

Arbache, J.S. (2001). "Trade Liberalisation and Labor Markets in Developing Countries: Theory and Evidence", *Institute of Applied Economic Research (IPEA)*, Rio de Janeiro, December. (*http://www.ipea.gov.br/pub/td/td_2001/td_0853.pdf*) [Accessed on July 20, 2007].

Arndt, C., P. Dorosh, M. Fontana, S. Zohir, M. El-Said and C. Lungren (2002). "Opportunities and challenges in agriculture and garments: A general equilibrium analysis of the Bangladesh economy", *TMD Discussion Paper No 107*. IFPRI, Washington DC.

Attanasio, O., P.K. Goldberg and N. Pavcnik (2003). "Trade Reforms and Wage Inequality in Colombia", *NBER Working Paper 9830*. Cambridge: MA.

Balassa, B. and Associates (eds.) (1971). *The Structure of Protection in Developing Countries.* Baltimore: John Hopkins University Press.

Baldwin, R.E. (2003). "Openness and Growth: What is the Empirical Relationship?", *NBER Working Paper* 9578. Cambridge : National Bureau of Economic Research.

Barro, R.J. (2000). "Inequality, Growth and Investment", *NBER Working Paper 7038*. Cambridge: National Bureau of Economic Research.

Behrman, J.R., N. Birdsall and M. Szekely (2003). "Economic policy and wage differentials in Latin America", *PIER Working Paper No. 01-048; Center for Global Development Working Paper No. 29*. April. (*http://ssrn.com/abstract=294517*) [Accessed on July 25, 2007].

Berman, E., J. Bound and Z. Griliches (1994). "Changes in the demand for skilled labor within US manufacturing: Evidence from the annual survey of manufacturers", *Quarterly Journal of Economics* 109(2): 367-397.

Bhagwati, J., D. Greenway and A. Panagariya (1998). "Trading preferentially: Theory and policy", *The Economic Journal* 108(449): 1128-1148.

Breusch, T. and A. Pagan (1980). "The Lagrange Multiplier test and its Applications to Model Specification in Econometrics", *Review of Economic Studies* 47: 239-253.

Bruno, G.S.F., A.M. Falzoni and R. Helg (2005). "Estimating a dynamic labour demand equation using small, unbalanced panels: An application to Italian manufacturing sectors", *mimeo*, (*http://www.stata.com/meeting/2italian/Bruno.pdf*) [Accessed on July 13, 2007].

———. (2004). "Measuring the effect of globalisation on labor demand elasticity: An empirical application to some OECD countries", *CESPRI Working Paper 153*. Universit`a Bocconi, Milano, Italy.

Carneiro, F.G. and J.S. Arbache (2003). "The impacts of trade on the Brazilian labor market: A CGE model approach", *World Development* 31(9): 1581-1595.

Chubrik, A., R. Giucci and I. Pelipas (2006). "Wage determinants in Belarus: Labor productivity and wage policy", Study conducted by the German Economic Team (GET) at the *Institute for Privatization and Management (IPM)*, Belarus. (*http://research.by/pdf/pp2006e04.pdf*) [Accessed on August 15, 2007].

Cline, W.R. (1997). *Trade and Income Distribution*. Washington DC: Institute for International Economics.

Coe, D. and E. Helpman (1995). "International r&d spillovers", *European Economic Review* 39: 859-887.

Collier, P. (1993). "Higgledy-piggledy liberalisation", *The World Economy* 16: 503-512.

Currie, J. and A. Harrison (1997). "Trade reform and labor market adjustment in Morocco", *Journal of Labor Economics* 15(3): S44-S72.

Deardorff, A. and D.S. Hakura (1994). "Trade and wages—What are the questions?", in J. Bhagwati and M.H. Kosters (eds.), *Trade and Wages—Levelling Wages Down?* Washington DC: AEI Press.

Dollar, D. and P. Collier (2001). *Globalisation, Growth and Poverty: Building an Inclusive World*. New York: Oxford University Press.

Dollar, D. and A. Kraay (2002). "Growth is good for the poor," *Journal of Economic Growth* (US) 7(3): 195–225.

Dollar, D. (1992). "Outward-oriented developing economies really do grow more rapidly: Evidence from 95 LDCs, 1976-85", *Economic Development and Cultural Change* 40(3): 523-44.

Dollar, D. and A. Kraay (2004). "Trade, growth and poverty", *Economic Journal* 114(493): 22-49.

Dornbusch, R. (1992). "The case for trade liberalisation in developing countries", *Journal of Economic Perspectives* 6(1): 69-85. Winter.

Duranton, G. (1999). "Globalisation, productive systems, and inequalities", *Paper presented at CEPR/CESPRI workshop on 'International Trade and Wage Inequality: Theory and Measurement'*, Milan, October.

Durevall, D. and F. Munshi (2006). "Trade liberalisation and wage inequality: Empirical evidence from Bangladesh", *Working Papers in Economics, nr 205*, Göteborg University. School of Business, Economics and Law, Department of Economics. (*http://www.handels.gu.se/epc/archive/00004857/01/gunwpe0205.pdf*) [Accessed on June 15, 2007].

Ebrill, L., J. Stotsky and R. Gropp (1999). "Revenue implications of trade liberalisation", *IMF Occasional Paper No. 180*. Washington DC: IMF.

Edwards, S. (1997). "Trade policy, growth and income inequality", *American Economic Review, Papers and Proceedings* 87(2): 205-210.

Ernst, C. (2005). "Trade liberalisation, export orientation and employment in Argentina, Brazil and Mexico", *Employment Strategy Department*, International Labour

Organization (ILO), Geneva. (*http://www3.ilo.org/public/english/employment/strat/download/esp2005-15.pdf*) [Accessed on July 3, 2007].

Fajnzylber, P. and W. Maloney (2001). "Labour demand and trade reform in Latin America", *World Bank Policy Research Working Paper No. 2491*. Washington D.C.: World Bank, January.

Falvey, R. and C.D. Kim (1992). "Timing and sequencing issues in trade liberalisation'', *Economic Journal* 102: 908-924, July.

Feenstra, R.C. and G.H. Hanson (1997). "Foreign direct investment and relative wages: Evidence from Mexico's maquiladoras", *Journal of International Economics* 42: 371-394.

Francois J. and C.R. Shiells (1994). *Modeling Trade Policy: Applied General Equilibrium Assessments of North American Free Trade.* Cambridge: MIT Press.

Fu, X. and V.N. Balasubramanyam (2005). "Export, foreign direct investment and employment: The case of China", *The World Economy* 28(4): 607-625.

Gourinchas, P.O. (1999). "Exchange rates do matter: French job reallocation and exchange rate turbulence 1984-1992", *European Economic Review* 43: 1279-1316.

Greenaway, D. (1993). "Liberalizing foreign trade through rose-tinted glasses", *Economic Journal* 103: 208-223.

Greenaway, D., R.C. Hine and P. Wright (1999). "An empirical assessment of the impact of trade on employment in the United Kingdom", *European Journal of Political Economy* 15(3): 485-500.

Greenaway, D., W. Morgan and P. Wright (1998). "Trade reform, adjustment and growth: What does the evidence tell us?", *Economic Journal* 108(450): 1547-1561, September.

Greenaway, D. and C. Milner (1993). *Trade and Industrial Policy in Developing Countries: A Manual of Policy Analysis*. London: The Macmillan Press Ltd.

Grossman, G. and E. Helpman (1991). *Innovation and Growth in the Global Economy.* Cambridge: MIT Press.

Hamermesh, D.S. (1993). *Labor Demand*. Princeton, NJ: Princeton Academic Press.

———. (1996). *Demand for Labor*. Columbia and Princeton: University Presses of California.

Harrison, A. (1996). "Openness and growth: A time-series, cross-country analysis for developing countries", *Journal of Development Economics* 48(2): 419-447.

Harrison, A. and A. Revenga (1995). "Factor markets and trade policy reform", *World Bank Manuscript*, 6/22.

Hausman, J. (1978). "Specification tests in econometrics", *Econometrica* 46(6): 1251-1271.

Herr, H. (2002). "Wages, employment and prices: An analysis of the relationship between wage level, wage structure, minimum wages and employment and prices", *Working Paper No. 15*, Berlin: Business Institute at the Berlin School of Economics. (*http://www.fhw-berlin.de/fhw2000/lehre_und_forschung/working_paper15.pdf*) [Accessed on July 25, 2007].

Hicks, J. (1932). *The Theory of Wages*. London: Macmillan.

Hoekman, B. and A. Winters (2005). "Trade and employment: Stylized facts and research findings", *DESA Working Paper No.7*. Department of Economic and Social Affairs. New York: United Nations.

Hoque, S. (2006). "Impact of trade liberalisation on employment in Bangladesh: A computable general equilibrium analysis", *Asian Business and Economics Research Unit Discussion Paper No. 26, Centre of Policy Studies*, Monash University. (*http://66.102.1.104/scholar?hl=en&lr=&q=cache:bj-T-v3aUkYJ:www.buseco.monash.edu.au/units/aberu/papers/working-papers-06/2306-serajul.pdf+*) [Accessed on August 23, 2007].

Hussain, M. I. and M. Rahman (1995). "Trends in the external sector: Trade and aid", in CPD (eds) *Experiences with Economic Reform: A Review of Bangladesh Development 1995*. Dhaka: Centre for Policy Dialogue. pp. 243-197.

ILO (2007). "Trade and employment: Challenges for policy research", *A joint study of the International Labour Office and the Secretariat of the World Trade Organization*, International Labour Office (ILO), Geneva, February 19.

IRBD (2002). *Employment and Labour Market Dynamics: A Review of Bangladesh's Development 2002*. Centre for Policy Dialogue: The University Press Limited.

Islam, R. (2007). "What kind of economic growth is Bangladesh attaining?", *Paper Presented at the International Conference on Development Prospects of Bangladesh: Emerging Challenges*, 2-3 December, 2007.(*http://www.bids-bd.org/images/rizwanul_islam.pdf*) [Accessed on August 23, 2007].

Jansen, M. (2003). "International trade and the position of European low-skilled labour" in Addison, J. and P. Welfens (eds.), *Labour Markets and Social Security: Issues and Policy Options in the U.S. and Europe*. Dordrecht: Springer.

Jansen, M. and A. Turrini (2004). "Job Creation, job destruction, and the international division of labor", *Review of International Economics* 12(3): 476-494.

Jean, S. (2000). "The effect of international trade on labour-demand elasticities: Intersectoral matters", *Review of International Economics* 8(3): 504-516.

Johnson, H.G. (1958). "Tariffs and the balance of payments: Comment", *The Quarterly Journal of Economics* 72(2): 301-302.

———. (1971). *Aspects of the Theory of Tariffs*. London: Allen and Unwin

Jones, R.W. (1965). "The structure of simple general equilibrium models", *The Journal of Political Economy* 73(6): 557-572, December.

Khan, F.C. (2000). "A decade of trade liberalisation: How has domestic industry fared in Bangladesh?", *Journal of Bangladesh Studies* 2(1): 31-37.

Khattry, B. (2003). "Trade liberalisation and the fiscal squeeze: Implications for public investment", *Development and Change* 34(3): 401-424.

Khattry, B. and J.M. Rao (2002). "Fiscal faux pas? An analysis of the revenue implications of trade liberalisation", *World Development* 30(8): 1431-1444.

Khondker, B. and M. Mujeri (2002). "Poverty implications of trade liberalization in Bangladesh: A general equilibrium approach", *Globalization and Poverty Programme*, UK: Department for International Development (DFID), (*http://www.gapresearch.org/production/MustafaPov.pdf*) [Accessed on August 15, 2007].

Kim, D. and R. Falvey (1992). "Timing and sequencing issues in trade liberalisation", *The Economic Journal* 102(413): 908-924.

Kirmanoglu, H. and M.E. Yazgan (1999), "Are real wages and competitiveness compatible in the Turkish manufacturing industry?", *Paper presented in the 19th Annual Meeting of Middle Eastern and North African Economies (MEEA) in conjunction with Allied Social Sciences Association in New York, New York, U.S.A.*, January 3-5, 1999. (*http://www.luc.edu/orgs/meea/volume1/kirmanog.html*) [Accessed on August 20, 2007].

Klein, M.W., S. Schuh and R.K. Tries (2003). "Job creation, job destruction and the real exchange rate", *Journal of International Economics* 59(2): 239-265.

Krishna, K.L. and G.S. Sahota (1991). "Technical efficiency in Bangladesh manufacturing industries", *Bangladesh Development Studies* 19(1-2): 89-106.

Krishna, P., D. Mitra and S. Chinoy (2001). "Trade liberalisation and labour demand elasticities: Evidence from Turkey", *Journal of International Economics* 55: 391-409.

Krugman, P.R. (2000). "Technology, trade and factor prices", *Journal of International Economics* 50: 51-71.

Lawrence, R.Z. and M.J. Slaughter (1993). "International trade and American wages in the 1980s: Giant sucking sounds or small hiccup?", *Brooking Papers on Economic Activity* 2: 161-226.

Leamer, E. (1995). "A trade economist's view of U.S. wages and globalisation", in Susan Collins (eds) *Imports, Exports and the American Worker*. Washington D.C: Brookings Institution.

Lee, E. (2005). "Trade liberalisation and employment", *DESA working paper, No. 5*, Economic and Social Affairs, October. (*http://www.un.org/esa/desa/papers/2005/wp5_2005.pdf*) [Accessed on August 16, 2007].

Lee, H. and D. Ronald-Holst (1995). "Regionalization and employment growth in the Pacific Basin", *Working Paper No. 94*, OECD Development Centre, February 1994. (*http://www.oecd.org/dataoecd/25/5/1919124.pdf*) [Accessed on July 7, 2007].

Levine, R. and D. Renelt (1992). "A sensitivity analysis of cross-country growth regressions," *The American Economic Review* 82(4): 942-963.

Ljunqvist, L. and T. Sargent (1998). "The European unemployment dilemma", *Journal of Political Economy* 106(3): 514-550.

———. (2005). "Jobs and unemployment in macroeconomic theory: A turbulence laboratory", *mimeo*. New York: New York University.

Loayza, N., P. Fajnzylber and C. Calderón (2004). "Economic growth in Latin America: Stylized facts, explanations, forecasts", *Central Bank of Chile Working Papers No.265*, (*http://www.bcentral.cl/estudios/documentos-trabajo/pdf/dtbc265.pdf*) [Accessed on August 16, 2007].

Lopez, R. and M. Schiff (1998). "Migration and the skill composition of the labor force: The impact of trade liberalisation in LDCs", *Canadian Journals of Economics* 31(2): 318-336.

Manasse, P. and A. Turrini (2001), "Trade, wages and "superstarts", *Journal of International Economics* 54: 97-117.

Matusz, S.J. (1996). "International trade, the division of labor and unemployment", *International Economic Review* 37(1): 71-85.

Matusz, S.J. and D. Tarr (1999). "Adjusting to trade policy reform", *Policy Research Working Paper 2142*, The World Bank, Development Research Group, Trade, July.

Melchior, A. (2006). "North-South trade and wages with complete specialisation", Paper presented to European Trade Study Group (ETSG), Vienna, September.

Milner, C. and O. Morrissey (1999). "Measuring trade liberalisation", in McGillivray, M. and Morrisey, O. (eds) *Evaluating Economic Liberalisation*. London: Macmillan.

Milner, C. and P. Wright (1998). "Modelling labour market adjustment to trade liberalisation in an industrializing economy", *Economic Journal* 108: 509-528, March.

Mouelhi, R.B.A. (2007). "Impact of trade liberalisation on firm's labor demand by skill: The case of Tunisian manufacturing", *Labour Economics* 14(3): 1-25. (*http://www.erf.org.eg/tenthconf/Labor_background/Rim_Mouelhi.pdf*) [Accessed on August 11, 2007].

Muqtada, M. (2003). "Promotion of employment and decent work in Bangladesh: Macro economic and labor policy consideration", *Employment Strategy Department*, International Labor Organization, Geneva, February 2003. (*http://www.ilo.org/public/english/employment/strat/download/bangldw.pdf*) [Accessed on July 10, 2007].

Mussa, M. (1978). "Dynamic adjustment in the Heckscher-Ohlin-Samuelson Model", *The Journal of Political Economy* 86(5): 775-791.

Nath, H.K. and K.A.Al Mamun (2005). "Trade, growth and inequality in Bangladesh: An empirical analysis", *Paper presented at the 75th Annual Meeting of the Southern Economics Association* in Washington D.C, November 18-20, 2005. (*http://www.shsu.edu/~eco_hkn/nath&mamun_tradegrowthinequality_nov2005.pdf*) [Accessed on August 2, 2007].

Osmani, S.R. (2005). "The impact of globalisation on poverty in Bangladesh", *Working Paper No. 65*, Policy Integration Department, National Policy Group, International Labour Office, Geneva, November. (*http://www.ilo.int/public/english/bureau/integration/download/publicat/4_3_413_wp-76.pdf*) [Accessed on July 25, 2007].

Panagariya, A. (1999). "Trade openness: Consequences for the elasticity of demand for labor and wage outcomes", *mimeo*, (*http://129.3.20.41/eps/it/papers/0308/0308007.pdf*) [Accessed on July 12, 2007].

Papageorgiou, D., A. Choksi and M. Michaely (1990). *Liberalisation of Foreign Trade in Developing Countries: The Lessons of Experience*. Washington DC: World Bank.

Polaski, S. (2006). *Winners and Losers: Impact of the Doha Development Round on Developing Countries*. Washington DC: Carnegie Endowment for International Peace.

Raihan, S. (2007). *Dynamics of Trade Liberalisation in Bangladesh: Analysis of Policies and Practices*. Dhaka: Pathak Shamabesh.

Rama, M. (1994). "The labor market and trade reform in manufacturing" in M. Connolly. and J. de Melo (eds.), *Effects of Protectionism on a Small Country: The Case of Uruguay*. Washington DC: World Bank.

Rashid, M.A. (2000). "Impact of policy reforms on industrial capacity and employment in Bangladesh", *Structural Adjustment Participatory Review Initiative (SAPRI)*, December 2000, (*http://www.saprin.org/bangladesh/research/bantrade.pdf*) [Accessed on July 15, 2007].

Razzaque, A., B.H. Khondker, N. Ahmed and M.K. Mujeri (2003). "Trade Liberalisation and Economic Growth: Empirical Evidence on Bangladesh.", *MIMAP-Bangladesh Focus Study No.03*. Dhaka: Bangladesh Institute of Development Studies.

Revenga, A. (1995). "Employment and Wage Effects of Trade Liberalisation: The Case of Mexican Manufacturing", *Policy Research Working Paper* 1524, Latin America and the Caribbean Country Department II, Country Operations Division 1. Washington DC: World Bank.

Robbins, D.J. (1996). "HOS Hits Facts: Facts Win; Evidence on Trade and Wages in the Developing World", *Development Discussion Paper* 557. Cambridge: Harvard Institute for International Development, Harvard University.

Rodríguez, F. and D. Rodrik (2001). "Trade Policy and Economic Growth: A Skeptic's guide to Cross-National Evidence", in Bernanke, B. and Rogoff, K.S. (eds.), *Macroeconomics Annual 2000*. Cambridge, MA: MIT Press.

Rodrik, D. (1997). *Has Globalisation Gone Too Far?*. Washington DC: Institute for International Economics.

———. (2002). "Institutions, integration and geography: In search of the deep determinants of economic growth". Available at *http://www.hks.harvard.edu/m-rcbg/research/d.rodrik_ksg_institutions.integration.and.geography.pdf*

Sachs, J. and A. Warner (1995). "Economic Reform and the Process of Global Integration", *Brookings Papers on Economic Activity*, 02/1995; 1(1995-1).

Sanchez-Paramo, C. and N. Schady (2003). "Off and Running? Technology, Trade, and the Rising Demand for Skilled Workers in Latin America", *Policy Research Working Paper* 3015. Washington DC: World Bank.

Scheve, K. and M.J. Slaughter (2004). "Economic Insecurity and the Globalisation of Production", *American Journal of Political Science* 48(4): 662-674.

Sen, K. (2003). "Globalisation and Employment in Bangladesh and Kenya", *Discussion paper 7, Globalisation Production and Poverty*. UK: Department of International Development (DFID).

Slaughter, M.J. (2000). "Trade and labor-market outcomes: What about developing countries?", *Conference paper, NBER Inter-American Seminar on Economics*.

———. (2001). "International trade and labor-demand elasticities", *Journal of International Economics* 54: 27-56.

Spilimbergo, A., J.L. Londoño and M. Székely (1999). "Income distribution, factor endowments, and trade openness", *Journal of Development Economics* 59(1): 77-101.

Solow, R.M. (1956). "A contribution to the theory of economic growth", *Quarterly Journal of Economics* 70: 65-94.

Stiglitz, J.E. and A. Charlton (2006). "Aid for trade", *A report prepared for the Commonwealth Secretariat*,(*http://works.bepress.com/cgi viewcontent.cgi?article=1008&context=joseph_stiglitz*) [Accessed on August 19, 2007].

Stolper, W. and P. Samuelson (1941). "Protection and Real Wages", *Review of Economic Studies* 9: 58-73, November.

Taslim, A. (2004). "Trade policy of Bangladesh: 1990-2003", *Article published in the Souvenir of the Dhaka University Economics Department Alumni Association (DUEDAA) Meeting*.

Tybout, J. (1990). "Making noisy data sing: Estimating production technologies in developing countries", *Journal of Econometrics* 53(103): 25-44.

Wacziarg, R. and K.H. Welch (2003). "Trade Liberalisation and Growth: New Evidence", *NBER Working Paper 10152, Cambridge: National Bureau of Economic Research.* (*http://ssrn.com/abstract=471941*) [Accessed on July 12, 2007].

Wei, S-J. and Y. Wu (2001). "Globalisation and inequality: Evidence from within China", *NBER Working Paper No.8611*. Cambridge: National Bureau of Economic Research.

Winters, L.A., N. McCulloch and A. McKay (2004). "Trade liberalisation and poverty: The evidence so far", *Journal of Economic Literature* 42: 72–115. March.

Wood, A. (1994). *North-South Trade, Employment and Inequality. Changing Fortunes in Skill-Driven World.* Oxford: Clarendon Press.

———. (1995). "How trade hurt unskilled workers", *Journal of Economic Perspectives* 9(3): 57-80, Summer.

———. (1997). "Openness and wage inequality in developing countries: The Latin American challenge to East Asian conventional wisdom", *World Bank Economic Review* 11: 33-57.

World Bank (1994). *Bangladesh: Privatization and Adjustment*, Dhaka.

———. (1999). "Bangladesh trade liberalisation: Its pace and impacts", *Poverty Reduction and Economic Management Division*, South Asia Region, November 23, 1999. (*http://siteresources.worldbank.org/INTSARREGTOPINTECOTRA/34004324-1120490724746/20926224/TradeLiberalizationItsPaceandImpacts.pdf*) [Accessed on July 25, 2007].

———. (2002). "Global economic prospects and the developing countries: Making trade work for the world's poor", *World Bank: Washington DC*, November 2001. (*http://siteresources.worldbank.org/INTGEP2002/Resources/00—FrtMatter—i-xxiv.pdf*) [Accessed on July 28, 2007].

———. (2004). "Trade policies in South Asia", *Report No. 29949.* Poverty Reduction and Economic Management Unit, South Asia Region, World Bank.

Zhu, S.C. and D. Trefler (2005). "Trade and inequality in developing countries: A general equilibrium analysis", *Journal of International Economics* 65(1): 21-48.

Index

K

L

M